About Island Press

Island Press, a nonprofit organization, publishes, markets, and distributes the most advanced thinking on the conservation of our natural resources—books about soil, land, water, forests, wildlife, and hazardous and toxic wastes. These books are practical tools used by public officials, business and industry leaders, natural resource managers, and concerned citizens working to solve both local and global resource problems.

Founded in 1978, Island Press reorganized in 1984 to meet the increasing demand for substantive books on all resource-related issues. Island Press publishes and distributes under its own imprint and offers these services to other nonprofit organizations.

Support for Island Press is provided by Apple Computers, Inc., Mary Reynolds Babcock Foundation, Geraldine R. Dodge Foundation, The Educational Foundation of America, The Charles Engelhard Foundation, The Ford Foundation, Glen Eagles Foundation, The George Gund Foundation, The William and Flora Hewlett Foundation, The Joyce Foundation, The John D. and Catherine T. MacArthur Foundation, The Andrew W. Mellon Foundation, The Joyce Mertz-Gilmore Foundation, The New-Land Foundation, The Jessie Smith Noyes Foundation, The J. N. Pew, Jr., Charitable Trust, Alida Rockefeller, The Rockefeller Brothers Fund, The Florence and John Schumann Foundation, The Tides Foundation, and individual donors.

THE
GLOBAL
CITIZEN

THE
GLOBAL
CITIZEN

Donella H. Meadows

ISLAND PRESS

WASHINGTON, DC □ COVELO, CALIFORNIA

The essays reproduced in this book have been edited and updated from works originally written for the *System Dynamics Review,* the Lebanon, NH, *Valley News,* and the *Los Angeles Times.*

Library of Congress Cataloging-in-Publication Data
Meadows, Donella H.
    The global citizen / Donella H. Meadows.
        p.   cm.
    Includes index.
    ISBN 1-55963-059-0 (alk. paper). — ISBN 1-55963-058-2 (pbk. : alk. paper)
        1. United States—Social conditions—1980–   2. Social problems.
    3. Human ecology.   4. Quality of life.   I. Title.
    HN65.M43   1991
    306'.0973—dc20   90-20899
                                                                            CIP

Printed on recycled, acid-free paper

Manufactured in the United States of America

10   9   8   7   6   5   4   3

THIS BOOK IS DEDICATED TO MY EDITORS AND TEACHERS:

Rick Minard and Marty Frank of the Lebanon, NH, *Valley News*

Guy MacMillan of the Keene, NH, *Sentinel*

Norman Runnion of the Brattleboro, VT, *Reformer*

Deane Wylie of the *Los Angeles Times*

Contents

THE
GLOBAL
CITIZEN

SYSTEM DYNAMICS MEETS THE PRESS

I T WAS 1985. Ronald Reagan had just been elected to a second term. The environment had disappeared as a subject of public discourse; people who lived in poverty were all welfare cheats who refused to get an honest job; the rest of the world was the backyard to which America crowed its perpetual "Morning." I couldn't stand it any more. I resigned my professorship in environmental studies at Dartmouth College to become a newspaper columnist.

I did it with no inside knowledge of the journalistic world. Until then I had met the press only as an object of reporting, and the meetings had been unsettling.

The press knew of me because the media find the field I work in— called system dynamics—fascinating. System dynamics is a set of techniques for thinking and computer modeling that helps its practitioners begin to understand complex systems—systems such as the human body or the national economy or the earth's climate. Systems tools help us keep track of multiple interconnections; they help us see things whole. Because much of conventional wisdom comes from seeing things in parts and focusing on one small part at a time, system dynamicists tend to have surprising points of view. They generate a lot of controversy. Hence the fascination of the press.

In 1969 I watched Jay Forrester (my mentor at MIT, the founder of system dynamics) try to explain to a nation in the midst of urban crisis why cities would be better off if governments pulled down public

1

housing instead of constructing it. As you might expect, that message infuriated city planners. The ensuing ruckus attracted the media like sharks to blood in the water.

By 1970 I was involved with a group at MIT making a system dynamics model of world population growth and economic growth. The press saw it as a global crystal ball, in which to foresee the future of everything. What an irresistible attraction! *Playboy*, of all publications, was the first to do an article about our work. There it was—an analysis of population growth, economic growth, pollution, resource depletion—right there among the naked ladies. A year or so later, when our book, *The Limits to Growth*, came out, we were given three whole minutes on the "Today" show to explain the growth, overshoot, and collapse of the world economy, just after a mouthwash commercial and just before a demonstration by the British dart-throwing champion.

From then on I watched the media misinterpret our book, label it a prophecy of doom, batter it, and discredit it. That was a painful experience, but one that led me to think long and deep about the crucial role of information and information-purveyors in the modern world.

My experiences with the media continued, sometimes funny, sometimes frustrating, occasionally fruitful. I kept coming back to the press because I thought my field provided valuable insights about the world. I wanted those insights to be spread widely—I knew they *must* be spread widely. System dynamics makes clear the overarching power of deep, socially shared ideas about the nature of the world. Out of those ideas arise our systems—government systems, economic systems, technical systems, family systems, environmental systems.

As Ralph Waldo Emerson once said,

Every nation and every man instantly surround themselves with a material apparatus which exactly corresponds to their state of thought. Observe how every truth and every error, each a thought of some man's mind, clothes itself with societies, houses, cities, language, ceremonies, newspapers. Observe . . . how timber, brick, lime and stone have flown into convenient shape, obedient to the master idea reigning in the minds of many persons. . . . It follows of course that the least change in the man will change his circumstances; the least enlargement of his ideas, the least mitigation of his feelings in respect to other men. If, for example, he could be inspired with a tender kindness to the souls of other men. . . . every degree of ascendance of this feeling would cause the most striking changes of external things.

So if we want to bring about the thoroughgoing restructuring of systems that is necessary to solve the world's gravest problems—poverty, pollution, and war—the first step is *thinking differently*. Everybody thinking differently. The whole society thinking differently. There is only one force in the modern world that can cause the entire public to think differently. That force is the mass media.

That was my reasoning when I set out to be a columnist. I was finding the state of the world and the feeble responses of policymakers intolerable. I didn't think that more writing for academics or preaching to the converted would help. I wanted to see a system-based, globally oriented, long-term viewpoint on the editorial pages of the newspapers. I kept waiting around for someone else to write it, but no one did. So I did.

I called the column "The Global Citizen" to emphasize the fact that my readers and I are part of an interconnected world system, whether we want to be or not. After five years of writing "The Global Citizen," I've learned a lot—about perceptions and paradigms, about the media, and about that wonderful public out there to whom we journalists try to speak. This book is a sample of what I've produced. This introduction is a summary of what I've learned.

THE PRESENT PARADIGM

A paradigm is not only an *assumption* about how things are; it is also a *commitment* to their being that way. There is an emotional investment in a paradigm because it defines one's world and oneself. A paradigm shapes language, thought, and perceptions—and systems. In social interactions, slogans, common sayings, the reigning paradigm of the society is repeated and reinforced over and over, many times a day. Whenever a speaker of an Indo-European language says a sentence, nouns and verbs reinforce the paradigmatic distinction between *things* and *processes* (in some other languages there are only processes). Every time you buy or sell something, you affirm a shared paradigm about the value of money. Every time the president rejoices when the gross national product (GNP) goes up, he strengthens the paradigm of economic growth as an unquestioned good.

Your paradigm is so intrinsic to your mental processes that you are hardly aware of its existence, until you try to communicate with someone with a different paradigm. Listen to an ecologist talk with an economist, a pro-lifer with a pro-choicer, a right-winger with a left-winger. In the difficulties of cross-paradigm discussion, both parties begin to be aware, often uncomfortably, of unspoken, fundamental assumptions they do not share.

System dynamicists were raised in the general culture, of course, long before they learned about system dynamics, so they are not uncomfortable in the normal paradigm of everyday life. But their systems training makes them very aware of the many unsystematic assumptions that permeate societal talk, political thinking, and daily news reports.

Here are a few of the common assumptions of the current social paradigm that seem to me to be clearly unsystematic and problematic. These are the assumptions that disturbed me enough to want to write a newspaper column:

- One cause produces one effect. There must be a single cause, for example, of acid rain, or cancer, or the greenhouse effect. All we need to do is discover and remove it.
- All growth is good—and possible. There are no effective limits to growth.
- There is an "away" to throw things to. When you have thrown something "away," it is gone.
- Technology can solve any problem that comes up. There is no cost to technology, no delay in attaining it, no confusion about what kind of technology is needed. Improvements will come through better technology, not better humanity.
- The future is to be predicted, not chosen or created. It happens to us; we do not shape it.
- A problem does not exist or is not serious until it can be measured.
- If something is "economic," it needs no further justification. E. F. Schumacher, writes, "Call a thing immoral or ugly, soul-destroying or a degradation of man, a peril to the peace of the world or to the well-being of future generations; as long as you have not shown it to be 'uneconomic,' you have not really questioned its right to exist, grow, and prosper."

- Relationships are linear, nondelayed, and continuous; there are no critical threshholds; feedback is accurate and timely; systems are manageable through simple cause-effect thinking.
- Results can be measured by effort expended—if you have spent more for weapons, you have more security; if you use more electricity, you are better off; if you spend more for schools, your children will be better educated.
- Nations are disconnected from one another, people are disconnected from nature, economic sectors can be developed independently from one another, some parts of a system can thrive while other parts suffer.
- Choices are either/or, not both/and.
- Possession of *things* is the source of happiness.
- Individuals cannot make any difference.
- People are basically bad, greedy, and not to be trusted. Good people and good actions are rare exceptions.
- The rational powers of human beings are superior to their intuitive powers or their moral powers.
- Present systems are tolerable and will not get much worse; alternative systems cannot help but be worse than the ones we've got.
- We know what we are doing.

I submit that the above statements are partially or wholly false, that they are implicit or explicit in virtually all public discourse, that they give rise to much of the counterproductive behavior of individuals and institutions, and that the harm done by them is incalculable. The only way I know to throw them into question is to question them, over and over, with as much documentation, clarity, and persuasiveness as possible, in the most visible public forums.

EVEN THE SIMPLEST SYSTEMS CONCEPTS HELP

The level of public discussion is so simpleminded that it doesn't take much to raise its quality. The most fundamental tenets of system dynamics can clear up significant muddles in public thinking.

Take, for example, the distinction between a stock and a flow—the distinction between the amount of water already in a bathtub and the amount pouring in through the faucet. I once wrote a whole column on the difference between the national deficit (a flow—the water pouring in—the rate at which we borrow) and the national debt (a stock—the water already in the bathtub—the accumulated debt). Reducing the deficit, I pointed out, will not reduce the level of debt; it will only mean that things are getting worse at a somewhat slower rate.

This is trivial systems theory, but I'm still not sure our politicians understand it.

It is a revelation to most people that you can increase the contents of a stock by reducing outflow as well as by increasing inflow, that, for example, economic wealth can be enhanced by repairing and maintaining old equipment as well as by investing in new equipment.

The effect of nonlinear relationships is not generally understood. For example, the public debate on the seriousness of soil erosion has yet to recognize that the relationship between soil depth and crop yield can be sharply nonlinear—that a little erosion may not have much effect, but a little more erosion may reduce output dramatically.

Other systems ideas that have immediate public relevance are:

- *Simple interconnectedness.* For example, energy conservation would not only save consumers money; it would also cut urban air pollution, acid rain, greenhouse gases, the production of radioactive wastes, the trade deficit, and defense costs in the Persian Gulf—only a few of the effects that would radiate through economic and environmental systems.
- *The astounding power of positive feedback and exponential growth.* Nigeria's population grew over the past thirty-five years from 43 million to 105 million. At the same rate of change over the next thirty-five years, Nigeria is expected to add another 207 million people, for a total of 312 million—43 million to 312 million over one human lifetime!
- *The time it takes for huge stocks to change.* After five years of *perestroika* the Soviet Union's economic situation has changed little. People are calling it a failure, not understanding how long it takes for a nation's capital plant, exhausted soils, and disaffected workforce to be revitalized.
- *The effect of delays on feedback.* Why oil prices went up and back down and why they will go up again.

- *The much greater importance of internal system structure than of triggering events.* One of the most controversial columns I ever wrote tried to divert attention from the immediate faults of Morton Thiokol's O-rings to the underlying structure that made a space shuttle accident almost inevitable.
- *The effect of bias in information streams.* Consistent Soviet and American overestimates of each other's weapons capability have been a major driving force in the positive feedback loop of the arms race.
- *The difference between information and physical quantities in systems.* I could write a column every week about the endless confusions between money and the real things money stands for.
- *How rational microbehavior can lead to disastrous macroresults.* The tragedy of the commons, the rise in malpractice insurance, economic cycles—there are hundreds of examples of this phenomenon. It is one of the most powerful concepts we have to offer because it *turns public discussion from the problem of blame to the problem of restructuring.*

Just one of these ideas at a time is enough to communicate in a newspaper column of 800 words. But they are ideas that can be communicated. Any systems concept, even a quite sophisticated one, can be expressed in words, in just a few paragraphs, as I hope the columns in this book demonstrate.

It's not easy to take on a social paradigm, and it's not welcome. When I started out, most newspaper editors did not think the environment belonged on the editorial page, much less anything directly attacking the most cherished beliefs of the society. I still have trouble getting editors to accept truly new ideas, especially ideas that attack the market system. My economic columns are the most unpopular ones I write, not with readers, but with editors. Many papers simply refuse to print them. You can't challenge the prevailing paradigm too directly.

But you can challenge it indirectly, bit by bit, again and again, presenting more and more evidence. Thomas Kuhn, who wrote *The Structure of Scientific Revolutions*, the seminal book about paradigms, says that what ultimately causes a paradigm to change is the accumulation of anomalies—observations that do not fit into and cannot be explained by the prevailing paradigm. The anomalies have to be presented over and over because there is a social determination not to see them.

Challenging a paradigm is not part-time work. It is not sufficient to make your point once and then blame the world for not getting it. The world has a vested interest in, a commitment to, not getting it. The point has to be made patiently and repeatedly, day after day after day. Fortunately, there are media like newspapers and television that have space to be filled day after day after day.

THE FILTERS

I have come to know at least fifty newspaper editors. They are well-informed people. They read four or five newspapers a day; editorial page editors read at least twenty opinion columns a day. They are disciplined, productive, and nimble with words. They make their deadlines every single day. Most of them follow a set of strong professional ethics about evidence, balance, truthfulness, and the public's right to know. Above all, they *care* about society and democracy and the information streams that hold a community or a nation together.

Like everyone else, however, they are embedded in a system whose structure, rewards, and punishments inevitably shape their behavior, not always for the good. The enterprises they work for put out a daily product on a rigid schedule that is not conducive to careful reflection. They are commercial enterprises that have to attract advertisers, appeal to the public taste, and make a profit. There is only so much space available every day, and competition for that space is intense.

Everything I've said about newspapers is even more true of the broadcast media. The result is a set of characteristics we are all familiar with—the standard and generally accurate set of criticisms about the media.

- They are event-oriented; they report only the surface of things, not the underlying structures.
- Their attention span is short, they create fads and drop them, they don't see slow, long-term phenomena (they ignored the greenhouse effect for decades until there was drought in the Midwest).
- They follow a herd instinct; they will send 1,500 reporters to one political convention, but no reporters will be on hand when crucial environmental policy is being made.

- They are attracted to personalities and authorities; they are uninterested in people they've never heard of.
- To meet time and space constraints, they simplify issues; they have little tolerance for uncertainty, ambiguity, tradeoffs, or complexity.
- They operate from skepticism; they have been lied to and manipulated so often that they don't believe anyone; they carry such a load of cynicism that they often unnerve sincere people who are telling the truth.
- They have a tendency to force the world to conform to their story rather than see the world as it is. (I have several times had the frustrating experience of being interviewed by a reporter who didn't want to hear facts that contradicted "the story.")
- They love controversy and think harmony is boring; they see the world as a set of win/lose, right/wrong situations; they are attracted to conflict and to things that aren't working; they do not pay attention to things that are working.
- They are strongly conservative; though they like to think of themselves as tough and uncompromising, in fact they challenge society only at its margins; most of the time, usually unconsciously, they reinforce the status quo and resist really new ideas.
- Also unconsciously they report through filters of helplessness, hopelessness, cynicism, passivity, and acceptance. They report problems, not solutions, obstacles, not opportunities. They systematically unempower themselves and their audience.

Why should anyone try to communicate messages of complexity, of structure, of long-term thinking, of inclusiveness, of empowerment through a system like this one? Because if we want a better world, we have no choice. And because it can be done, in spite of that negative list I've just made. I've learned that communicating through the media is harder than I thought, but also more possible and rewarding than I thought.

IT CAN BE DONE

My greatest help has been a handful of editors and television producers who have taken me in hand, coached me, and criticized me. Slowly they have taught me to stop resisting the strictures of the media and to work within them, without, I hope, losing my purpose or message.

My greatest problem at the beginning was keeping my columns under 800 words (the earliest columns in this collection are the longest ones). One of my editors thundered at me, "George Will can write less than 800 words. Mary McGrory can write less than 800 words. Why can't *you* write less than 800 words?" Another reminded me that I didn't have to say everything all at once. With a weekly column, I'd always have another chance.

Be clear, not fancy, they told me. Use everyday language. Be specific, not abstract. Offer easily imaginable examples. Be sure your words make pictures in people's heads. Be sure the pictures are the ones you intend.

Use most of your column for the evidence, they said. Tell stories, give statistics, show the impact of the problem or the solution on the real world. People can form their own conclusions if you give them the evidence. Don't take much space for grand, abstract conclusions; let the reader form the conclusions.

Use a hook to the news—that point was hard for an academic like me to get. If you're writing about energy conservation, tie it to the shooting down of a commercial airliner over the Persian Gulf. If you're writing about the ozone hole, point out that the Senate just ratified a treaty to combat it. People have to know that what they're about to read is important. They think the daily news is important. So use that hook, even if you're not going to talk about the daily news.

Write an interesting lead. Another editor once blasted me with, "That was the most terrific column you ever wrote, but it had a boring, killer lead." A killer lead is an opening sentence that makes the reader yawn and turn to the sports page.

Never write in an apologetic tone, they told me, or a defensive one. Never, ever, ever, condescend to the reader. Never present a problem without providing at least a hint of what to do about it. Don't get people all riled up and then drop them into helplessness.

A television producer taught me an important lesson—whatever your story, tell it through *people*. Human beings are much more interested in other human beings than in ideas. Don't shy away from personalities, don't try to hide your own personality (difficult for a scientist who has been trained otherwise). If you care about something, let your care show as well as your evidence. If you're writing about someone else who cares, or who doesn't, make that person as real and whole on paper as you possibly can.

Be humble. You don't know everything. Even system dynamicists don't know everything. In fact no human being knows much of anything, compared with the immense wonders and uncertainties of the universe. So keep a sense of perspective. Say what you can say and no more; say it with the appropriate degree of certainty and no more. *That* is a hard lesson to follow. It's a torture every day and a duty, a discipline and a Zen koan, the bane of my existence and the best challenge of my life.

You can decide for yourself, reading the columns collected here, how well I have learned these lessons. You will also note a few lessons I have imposed upon myself. Though I often despair about the state of the world, I also tell good-news stories. That's uncharacteristic for a columnist, but necessary for a reformer who wants to show that a better world is both imaginable and feasible. I often show up personally in my columns, a practice much frowned upon in the profession, but one that makes me feel more honest. I don't think columns should sound like they come from God. I like to be reminded that authors are always limited, biased, quirky human beings.

SO WHY A BOOK?

Because my readers kept asking me for one. Because Island Press came along with the suggestion that I do it. Because the issues I write about do not go away with the speed of the fast-fading contents of the daily newspaper. And because, as I discovered as I waded into five years of accumulated columns, my little squibs stand together perhaps better than they stood apart.

Though I was taking on energy one week and the greenhouse effect the next and the absence of political leadership the week after that, in fact those three topics fit together in a way that a single 800-word column couldn't express. Whatever the weekly topic, my columns have all flowed from a systems-trained mind that does its best to operate with a holistic picture of how things are integrated. It was fun to put the pieces back together again.

THE REWARDS

As my columns appear in more papers and reach more people, I hear from some of those people, and that is the gratifying part of this exercise. I get letters and phone calls, sometimes angry, sometimes crazy, but mostly thoughtful, appreciative, supportive, and interesting.

People send me additional material about a subject I've written about. They tell me about steps they are taking to correct a problem. They point out my mistakes, usually very patiently. They ask questions and suggest column ideas. They let me know when they think one of my columns is below standard, and they're always right. They tell me they've cut out one of my pieces and sent it to their senator or brother-in-law, or they've read it to their ninth-grade class, or they've stuck it up on their bulletin board at work.

There's a great audience of engaged, active global citizens out there, yearning to make sense of their world and to make that world better. They put ideas to work. They are the living receptacles, perpetuators, and changers of the paradigms of society. They—you—are the key to a sufficient, sustainable, fair, and wonderful future.

PERSONAL
NOTES

IT'S NOT usually possible to tell, when reading "The Global Citizen," that the column is written by an ex-biophysicist and college professor who lives on a sheep farm and writes on a Macintosh computer with a Buddha sitting on top of it, a cat on her lap, and a dog at her feet.

Those personal facts are supposed to be irrelevant to the logic and point of view of the column. They aren't, of course. So every now and then I let the farm into the writing, usually when it is so much on my mind that I have no choice.

In addition to the farm and system dynamics and my basic science training, the other formative experience that shapes my writing is *The Limits to Growth*, the book about the computerized world model, which generated so much controversy and misunderstanding after it was published in 1972. For reasons I can't explain, I have written only once directly about that book and its message. Since it is very much a part of me, I have also included that column in this personal section.

Living Lightly and Inconsistently on the Land

I WAS RAISED in Illinois, as a good red-blooded American kid, eating Jello, white bread, canned peas, and Midwestern steaks. I watched Howdy Doody and played softball and canasta. On my sixteenth birthday my father gave me a decrepit old car. I used it to drive to my summer job in a drugstore at a shopping mall. Everyone I knew lived just like me. I didn't know there was such a thing as a lifestyle.

I went to college on a scholarship and developed tastes for things that went beyond my family's ken—artichokes and opera and Shakespeare. As a chemistry major I did a term paper on chemical additives in food, and for the first time I began to make consumer decisions that didn't come from habit. I read labels and tried to buy foods that were mainly composed of food.

Getting married and moving east didn't induce many changes until I started studying biochemistry. As I learned more about the body's chemical processes, I started putting more whole wheat flour in things I baked, using less sugar and fat, and serving more green vegetables. I eliminated the Jello and everything in cans and swore off soft drinks and coffee. All this was done for our own health, not from any sense of global responsibility.

The quantum leap in lifestyle came when my husband and I spent a year driving through Turkey, Iran, Afghanistan, Pakistan, and India (in those days Americans were welcome in all those countries). In India we became vegetarians because it was difficult to find meat. In the Muslim countries we couldn't buy alcohol. Our clothing had to be simple and practical. Hot showers became major luxuries. We were cut off from television, radio, and even newspapers for weeks at a time.

Mostly we lived as the villagers around us did, and we discovered that we were perfectly happy to do so.

Coming home was a shock. Looking with Asian eyes, we couldn't believe how much *stuff* people had. We saw how little the stuff had to do with happiness. We also had strong memories of the poverty, the erosion, the deforestation, and the hunger we had seen. The world was very real to us. We resolved to live our lives in a way more consistent with the whole of it.

At first we had little desire for material things. That wore off, of course, and we became Americans again. But we kept our life simple. We continued to be vegetarians. We traveled by mass transit. We made our own clothes and bread and even furniture. We asked a long set of questions about everything we bought. Is this spinach organic or raised with pesticides? Were these bananas grown on an exploitive plantation or in a worker-owned cooperative? Is our electricity from a hydro dam or a nuclear power plant? If we buy plastic bags, how many nasty chemicals have we caused to be released somewhere? Can we get along without plastic bags?

We were the best global citizens we knew how to be. And we were a pain in the neck. We regarded most of the people around us as unaware, unconsciously wreaking planetary destruction for short-term gratification. We separated ourselves from them. It didn't occur to us that setting up us/them and right/wrong categories might be the surest way of all to wreak planetary destruction.

We moved to New Hampshire because we wanted to restore a beat-up farm to ecological health and to live more self-sufficiently. We ripped the house apart and put it back together with proper insulation. We added space so six or eight people could share the place without stifling one another, and since then we've lived communally. We heat the house and our water partly with wood, partly with oil. We cut the wood with a chain saw, split it with a hydraulic splitter attached to a tractor. We cook on a woodstove sometimes, but an electric stove mostly. Our electricity comes partially from a nuclear power plant.

We grow nearly all our vegetables, all our eggs, some fruit. We grow organically, of course. We've made the soil much better than it was when we came here. We dry, can, freeze, and pickle enough to get through the winter. We buy milk from another farm, grain from the co-op, and ice cream from the supermarket. We raise sheep for wool, some of which we sell, some of which we spin, dye, and knit our-selves. We sell the meat. We recycle organic garbage to the chickens, cans and bottles to a recycling center. We wash out and reuse plastic bags; we use old newspapers to start fires in the woodstoves. We put tons of junk mail out in the weekly garbage pickup, which goes to a trash-to-energy incinerator.

We bought a television when the Red Sox were in the World Series. It's hardly been on since, but classical music plays all day on a CD player. We have old-fashioned spinning wheels and modern com-puters, an energy-efficient Honda and a wildly inefficient Dodge pickup truck for farm work. We travel by jet all over the world to do environmental work, probably burning up 100 times as much fuel as any other family in our town.

I assume that the inconsistencies in this "lifestyle" are obvious to you. We try to live lightly on the land in a culture where that's impossible. But we have lightened up about our own compromises and those of others. We do our best, we're always willing to try to do better, and we're still major transgressors on the ecosystems and resources of the planet. We're a lot more tolerant of our fellow trans-gressors than we used to be.

As a child in the middle-class Midwest, I lived out of a subconscious sense of *abundance*. That sense permits security, innovation, generosity, and joy. But it can also harbor insensitivity, greed, and waste. After returning from India, I lived out of a sense of *scarcity*. That is fine when it fosters stewardship, simplicity, and frugality, but not when it leads to grimness, intolerance, and separation from one's fellows. Now I try to base my life on the idea of *sufficiency*—there is just enough of everything for everyone and not one bit more. There is enough for generosity but not waste, enough for security but not hoarding. Or, as Gandhi said, enough for everyone's need, but not for everyone's greed.

Lambing 101

EVERY YEAR on the first of September we separate the rams from the ewes for two months, to postpone the lambing until the first warm days of April. I now know that when we did that last fall, it was a week too late. Last week we found newborn twins on the January snow. Lambs have been coming ever since.

From the beginning of our relationship with sheep, they have been doing that to me: surprising me and showing me how little I really know.

I started with three ewes in the summer of 1977. For months all was well because they didn't have to do anything but eat and I didn't have to do anything but feed them. But as the winter wore on and lambing approached, I knew that the real test of the shepherd was just ahead. I read every book I could find on sheep obstetrics. I hung around old-timers, asking questions. I watched some training movies. I became an expert on malpresentations, milk sickness, tetanus, and mastitis. Upon request I could deliver a scholarly lecture on the theory and practice of lambing. What I couldn't deliver, it turned out, was a lamb.

In the fullness of time a call came to me at my job at the university. "There's a big bag of liquid like a balloon coming out of Clove. Is that what's supposed to happen?" None of my books had mentioned big bags of liquid. I sped home, envisioning prolapses and breech deliveries. By the time I got there, Clove was calmly licking her twins and supervising their first wobbly attempts to stand.

I missed the second lambing too. I came out in the morning to find another set of twins already up and nursing. Then came the third lambing, the real humiliator. I was alone that day and came into the barnyard to find the head of a lamb emerging, but no hooves. My textbooks informed me that this was a serious malpresentation. That lamb was never going to get out of there without my help.

The books never hinted, however, that the ewe might not, at such a delicate moment, want any help. Not a single book had said, "First, catch ewe" or warned that this might be an arduous task.

Half an hour later I had the ewe in the pen, both of us breathless, and I had fetched the requisite bucket of hot soapy water. "Reach in and straighten out the forelegs," said the book. "Then deliver as usual." I discovered that there were several deficiencies in this plan. First, one cannot reach into a space in which a lamb is stuck. There is not even room for two fingers, much less a hand. Second, additional pressure in that tight enclosure caused the ewe to strain and eject the intrusion with bone-crunching force. Third, when I finally maneuvered the lamb's head back in far enough to make room for my hand, the space inside was hot, slippery, surging, and jam-packed with little legs, only two of which were the ones I was seeking.

I shall draw a merciful curtain over the rest of the scene and say only that it finally came to an end, with the ewe alive but not the lamb. Though I actually knew much more by then, I felt as if I knew nothing. My expertise had been put in its place. The reality of lambing was nothing like what I had built up in my head from all that secondhand knowledge. I had created for myself an orderly, antiseptic, shallow picture of a messy process that had smells, feels, sounds, emotions, and mysteries I could never have imagined.

What finally got that lamb out was a little bit of book learning and a lot of desperation, moment-by-moment sensing, a willingness to shut down the yammering of my mind and surrender to some wordless inner wisdom that guided the fingers and the legs and the forces and counterforces. That inner wisdom has helped me deliver every difficult lamb since then. All my mistakes have come from failing to follow it.

That lesson, and many others the sheep have administered over the years, will seem obvious to farmers and others who work directly with nature. Such people do not have the arrogance of the book-learned, who forget that words are not experience, that categories are only in the mind, and that knowing is not the same as doing. If such people ever had that arrogance, nature knocked it out of them long ago. I, who work too much with words, have to be reminded regularly. It's one of my reasons for keeping sheep.

To me the ideal leader of a government or corporation or any human organization is someone who has both learning from books and humility from experience—the kind of experience when you're right in the middle of things, everything depends on you, and you will be informed quickly and mercilessly about your mistakes. That leader would be at home with the knowledge our society has compiled, but would also be aware of the deep ignorance that remains. He or she would have more than theories about the complex realities of raising food, or being unemployed, or pulling coal from the earth, or being a soldier in a war, or bearing a child.

My ideal leader would proceed experimentally and tentatively, immersed in the moment, listening carefully to both inner and outer wisdoms, alert for signals that something is not working. He or she would be able to learn, to change course, to admit being wrong.

We seldom have leaders like that because we don't choose them. We prefer people of full-steam-ahead certainty. Maybe that's because we're so aware of and uncomfortable with our own uncertainties that we need to think that someone, somewhere, has the situation under control. Or maybe we don't keep enough sheep, children, students, or equivalent lesson-givers to keep us aware of our fundamental uncertainties and of the deep mysteries of the world.

A Time of Death and Life

THE SHEPHERD'S year ends and begins in November. That's when we bring the sheep up from the pasture, butcher last spring's lambs, and turn the ram in with the ewes to make lambs again. The record book closes on one cycle, and the next cycle begins.

I think of it as The Time When We Haul Sheep Around.

This year the first haul was a ram trade. Our ram Polka was born on our farm, and his mother, sisters, and daughters made up most of our flock. Too much inbreeding; Polka had to go. So we arranged a trade with a neighbor who also had a handsome ram who had been around too long.

We were sorry to lose Polka because he was the only ram we'd ever had who wasn't a complete nuisance. Rams generally treat every moving object except ewes as something to be knocked over—that's where the verb "to ram" came from and why "The Rams" is a perfect name for a football team. A ram can weigh 250 pounds and prefers to come at you on the run, head down. The rule of the barnyard is: Never turn your back on a ram.

Polka was an exception. The new ram we traded him for wasn't. He was a 100 percent, all-American ram. We had to go back to using evasive bullfight maneuvers to do our chores. The sheep corral was his, and he defended it against all comers. We named him Caspar Weinberger.

We hauled in another new sheep, a yearling ewe, from a different neighbor. We named her Faith, to go with the two ewe lambs we kept from our own flock—Hope and Charity. We thought Faith, Hope, and Charity would balance Caspar Weinberger very nicely.

Because we had added three young ewes, we had to get rid of some older ones. The flock was at the carrying capacity of the pasture. After much agony and long perusal of the records of lambings and fleece weights, we decided that Vanilla and Angelica were the ones to go. We hauled them to the Thetford Auction. We didn't stick around to see who bought them; we prefer to keep auction outcomes ambiguous. They were fat, well-formed ewes with good years of breeding still left in them; presumably someone recognized that.

The final haul was the five ram lambs we took to Sharon Beef, the slaughterhouse. This is an annual trip, which we do with crisp routine and a certain serenity. We unloaded the rams into clean stalls at Sharon Beef on a Sunday night. The next Friday we picked up wrapped, frozen, labeled lamb meat to deliver to the freezers of our customers. We also picked up the pelts to scrape, salt, and cure to become sheepskin rugs.

The slaughterhouse is meticulously clean and efficient, and all the business is done with neighbors. Neighbors do the butchering and neighbors buy the meat and rugs. The income arrives just before the December tax bill, when we most need it.

When I talk happily about this time of year, some folks, nearly always meat eaters, ask how I can be so cruel as to take my lambs—in whose births I assisted, whose growth I've overseen, whose mothers I call by name—to slaughter. It's a question that can only come from an urbanized culture like ours, where most people live a long way from the sources of their food.

You don't have to live on a farm very long before you come to terms with life and death, with the Novembers when you kill the lambs and start the lambs. You don't become hard or unfeeling; rather you become accepting. You know that birth and death are not separable and that deaths are necessary, so that the ratios of rams and ewes and sheep and pastures will be right, and so there will be meat to feed people. On a farm every stage of the cycle—breeding, birth, growth, maturity, death—has beauty and dignity.

November isn't the exciting high of spring when the lambs are born and the daffodils bloom. It's the time of preparation for spring. The dead-looking daffodil bulbs go into the ground, and the ram goes in with the ewes. The fall is the time to remember that nature turns death into new life. The garden takes last year's cornstalks and fallen leaves and sheep manure and turns them into next year's tomatoes and broccoli. The sheep turn last year's hay into next year's wool and lambs. And who knows what tasks and achievements, joys and sorrows, our customers will produce out of the energy from that lamb meat.

It was Gandhi who pointed out that in spite of all the death in the world, what persists is life.

Thoughts While Cleaning the Living Room

MY FEMALE friends throughout the world have been sending me thought-provoking statistics about women's work.

Twenty years ago only 10 percent of the women in the industrialized countries worked full time. Now over 50 percent do. But the average amount of time men spend on housework has not increased.

In Norway the more children in a family, the less total time the father spends with them. In Canada the more children in the family the more time the man spends watching television.

In the United States the income gap between full-time working men and women has not narrowed. Only one working woman in four earns a wage high enough to support herself and her family.

Nearly all Russian women work full time, many in traditional male roles, from heavy construction to high bureaucracy. Soviet men do almost no cooking, cleaning, laundry, shopping, or childcare. Doctors in the U.S.S.R. are among the most poorly paid and least respected workers, and nearly all of them are women.

These figures make me sad. They say not only that the power relationships between men and women are still badly skewed, but also that industrialized society has not yet learned to value the crucial tasks of nurturing people and maintaining homes.

I suppose everyone grows up as I did, certain that there is no worse way to spend your time than doing dishes or scrubbing floors. Being forced to do these things as a child doesn't help. As a female child, though, I absorbed without question the idea that these jobs were mine and that I would do them all my life.

It took years before I finally shook off the childhood associations and noticed how I actually feel about household chores. I don't mind them. Work is not a bad thing—it is an integral part of a good life. Housework is as pleasant as any other. Vacuuming the living room is healthful exercise, and it produces economic benefit (maintaining the worth of capital assets) and aesthetic benefit (a clean living room). On all those counts it beats sitting through a faculty meeting or writing a column.

I have gradually learned that an ongoing responsibility for a house or a child is an opportunity to practice fine human virtues—selflessness, patience, practicality, orderliness, intuition, love—womanly qualities not because of genetics but because of the way most women spend their lives.

I've also come to realize that nurturing and maintenance hold the world together. Their social worth is inestimable. The women of Iceland made that point dramatically one day in 1975. All the women in the country, paid and unpaid, took a day off and gathered together for a discussion of women's rights. In the words of one observer, "The wheels of society came to a screeching halt, and no one questioned the value of women's work again."

Why does anyone ever question the value of "women's work"? Why don't we honor it? I mean really honor it, with decent wages and personal respect? At the bottom of the ladder of compensation and prestige are people who clean up or care for people—homemakers, nurses, attendants at day-care centers or retirement homes, social workers, maids, launderers. Most of them are women or minorities. Above them are people who care for machines, still higher people who care for flows of paper, highest of all people who care for money. Where did we get that set of priorities?

Either we valued cleaning and caring so little that we foisted them off on people we didn't respect, or we thought so little of women that their jobs became devalued by association. However it happened, we now take jobs that are pleasant enough within the varied routine of a household—peeling potatoes, wiping the noses of two-year-olds, serving food to the table—and make some people do them full time. In the process the jobs become boring and demeaning. We look down on the people who do them, and all but the strongest of those people look down on themselves.

The economist Kenneth Boulding once said, in response to a forecast that someday every American would be earning $100,000 a year, "So what? Someone will still have to take out the garbage." Whoever that is, no matter how much she or he makes, that person is not likely to be respected or have self-respect, as long as we view cleaning and caring as distasteful. Around those jobs we will continue to construct social classes:

- Upper-class people never clean up after themselves or care for the physical needs of others. They consider their time too important to spend that way.
- Middle-class people clean up after and care for themselves and a few others in reciprocal arrangements—I'll do your laundry if you'll shovel the snow—and they consider themselves just ordinary folks,
- Lower-class people clean up after themselves and others all day, every day, and figure that's all they're good for.

As a woman I don't want to be liberated out of the middle class. I want everyone to be liberated into it. Especially those men who don't help their working wives, don't respect the work those wives do, and don't realize what they're missing by never vacuuming the living room.

A Lament from the Countryside

A

S I WRITE, the sun, low in the west, is shining out from beneath storm clouds. This is the third bank of clouds to come over this month, drop ten minutes of soft rain, and pass on. A few minutes ago I went out and scratched in the garden with a hoe. Under the damp surface the soil is dry, dry, dry.

It's bad enough to watch the pasture turn brown and the streams drop and to think it's just a bad year. It's much worse to think that this drought may be attributable to human causes. No one is sure that the weird weather this hot summer of 1988 is due to global warming from the greenhouse effect. But it could be, and the possibility is profoundly disturbing. What we used to call acts of God may now in fact be acts of man.

Here in New England we're used to icy, snowy, hot, dry, windy, or wet mayhem. Nature is variable and sometimes violent. We Yankees have developed a full armament of curses, prayers, patience, and ingenuity to deal with it.

But until now, whatever happened, we knew that natural cycles would eventually turn and make things right again. We never had to consider the possibility of permanent breakdown. And we never had to think of ourselves and our fellow human beings as perpetrators. We had to put up with wild weather, but not with blame or guilt.

Of course when the asparagus came up out of the ground all crooked, I used to say it was because of the nuclear plant down at Vernon. But that was a joke. What's happening now is no joke.

The maple trees look terrible. Whole branches are dead. It's pear thrips, the experts say. Just another cycle, like the spruce budworm that defoliates the forest and then goes away. But it could also be that the trees are stressed by acid rain (tree rings show that their growth has slowed in the last twenty years). Maybe the pear thrips are just opportunists, moving in to polish off an already sickened forest. If so, we have a real score to settle with the coal burners of Ohio—and with our own oil furnaces and car exhaust.

The experts are not sure what's wrong with the maples. They won't be sure for years. Meanwhile, every time I look at them, I feel a sense of creeping disaster.

When I moved here sixteen years ago, the whippoorwills drove me crazy every June with their night calls. I haven't heard one in years. Friends who monitor nesting birds for the Audubon Society tell me there's a significant drop-off in warblers, thrushes, tanagers, and most other migrants, most likely because of the destruction of tropical forests down south where the birds winter.

Again, the experts aren't sure. The Fish and Wildlife Service says it'll be awhile before we can even be certain the downward trend is real. And then to sort out its causes over thousands of miles of migration routes may be impossible.

No rain. No maple trees. No songbirds. Is that the world we're coming to? There are more than three times as many people as there were in 1900. The number of machines and factories has multiplied at least twentyfold. The rate of fossil fuel burning has increased by a factor of ten, producing a sixfold greater stream of sulfur dioxide emissions and tenfold greater emissions of nitrogen oxide and carbon dioxide. Hence the acid rain, the global warming, and perhaps the world's first human-induced drought. Was all this growth worth the price?

In the drying countryside, where nature used to be a solace, every little abnormality now generates dark thoughts. How come the bean leaves look so funny? Why is there such an outbreak of slugs? What will go wrong next? If human beings are unbalancing the atmosphere, changing the climate, acidifying the rain, eliminating whole species of life—rivaling acts of God and then becoming victims of those acts— where do we stand in the scheme of things? We have godlike power, but we ourselves suffer from that power, along with the rest of nature.

C. S. Lewis once wrote, "What we call Man's power over Nature turns out to be a power exercised by some men over other men with Nature as its instrument." The tremendous forces we have learned to extract from coal, oil, and gas have become, through the mediation of nature, forces exercised by city folk over country folk, by rich people over poor people, and by this generation over all future generations.

I'd give anything to be kept awake again by a whippoorwill, just to have some assurance that all's right with the world.

Missing More Than a Home

LAST CHRISTMAS I was homeless, in a way, for awhile. I *felt* homeless anyway, and the pain was sharp enough to give me a small insight—perhaps as much as we who have never really lacked a home can have—of what homelessness means and why it demands a larger response than building shelters.

In fact I had a welcoming roof over my head. I was well fed. I was within reach of family and friends. Those who are actually, chronically homeless would have traded places with me in a minute.

But I had left my home of sixteen years, and I didn't expect to return. I was staying in someone else's house, much nicer than my own. But not my own.

I had planned to serve at a community Christmas dinner for the less fortunate. You know the kind—lots of turkey in a church basement, metal folding chairs and cheap tinsel, and a singing of carols after the meal. I wonder if the recipients of those dinners realize what a gift they are giving the donors. For years I had been the dispenser of good things, the baker of cookies, the server of large, festive dinners. I badly needed to fill that role on Christmas Day. If I couldn't, I wouldn't quite know who I was.

The community dinner was canceled, and there I was, without an identity. I dropped in at friends' houses. I received their kindness with nothing to give. I was miserable. I did a lot of meditating about what that word "home" means.

In economic terms home is a place to accumulate the physical things that support you in leading a productive life. Home is where you wash clothes, cook food, and restore yourself to go out and do battle with the world.

Most of us never think how impossible it is to get along without a simple place to keep things. Street dwellers in India pay up to half their meager incomes for firewood to burn in open cooking fires. Helping agencies once tried to give them small, portable stoves that use half as much fuel, but there was no place to store them. The stoves were all stolen.

Home is the one place you can arrange to suit your own pleasure. It's where you keep paintings or pets or hobbies. Home can be an outward expression of your inner landscape, a statement to the world of who you think you are, or who you'd like to be.

That may seem a trivial function, but the most decrepit shanties in the worst slums have green plants growing out of tin cans or colorful magazine pictures on the wall. Some touch of beauty and cheer and self-expression, however humble, is a human need nearly as basic as shelter from the rain. Where do you put your pictures when you don't have a wall?

Home is where you can be exactly who you are without apology. Those who live with you may not approve in every detail, but at least they're not surprised. They're used to you. That's a comfort beyond price. Imagine being on display before strangers on the street every hour of the day, every day.

Home is a place you can share with friends. Without a home you can't be hospitable. Those who wander the streets exhibit their own kinds of generosity, but their repertoire for sharing is limited. They can't have the small pleasure of inviting someone in for a beer or a cup of tea.

This Christmas, with joy and gratitude, I am back home, and I am left with the understanding that homelessness deprives people of more than shelter. It deprives them of a place from which to be productive and giving, to be restored, to be welcomed, to be themselves, to give physical expression to their personalities. The homeless are dispossessed of a part of their humanity.

I have mused often over the words of Mother Theresa: "Much of the suffering in the world is caused because of want of food, want of clothes, but it is caused even more because of want of love. Many people are not only naked for a piece of cloth, they are naked without human dignity that has been stolen from them. Homelessness is not only being without a house, but is also being dejected, unwanted, unloved, uncared for. The greatest injustice we have done to our poor people is that we have forgotten to treat them with respect, with dignity, as children of God."

The homeless are not all easy to treat with respect. Some of them surrendered their dignity long ago to alcohol, drugs, dementia, or impenetrable hostility. But then again some of them are children. Mother Theresa would scoop them all up into her undiscriminating love, but she's a saint. I'm no saint, and I don't know the solution for homelessness. What I do know is that it requires more than warehouses to remove the homeless from our sight. It requires attention, engagement, the willingness to focus on homeless people not as statistics but as individuals who need to be welcomed back as full members of the human race.

The Limits to Growth Revisited

A S ENVIRONMENTAL problems make the headlines again, people who remember *The Limits to Growth* have been asking me, "Were your predictions right? Ozone hole, acid rain, greenhouse effect—is this it? Are these the limits to growth?"

I suppose if they wanted to hear the answer "no," they would ask someone else.

In 1972 I coauthored *The Limits to Growth* as a report to the Club of Rome. The book was based on a computer model of global population and economic growth. It said, in essence, that the kinds and rates of physical growth the modern world is accustomed to cannot go on forever, or even very much longer.

That message raised a surprising ruckus. We were attacked from the left and the right and the middle. The book was banned in the Soviet Union and investigated by President Nixon's staff. The Mobil Corporation ran ads saying "growth is not a four-letter word." Disciples of Lyndon LaRouche and the National Labor Caucus picketed our public appearances. Mainstream economists competed with one another to see who could write the most scathing reviews.

Seventeen years later, in 1989, the world population had risen from 3.6 billion to over 5 billion. Tropical forests were being cut down at a rate of 26 million acres per year (an area the size of Pennsylvania). The rate of fossil fuel burning was 40 percent higher than it was seventeen years before, and the carbon dioxide released from that burning was deranging the chemical composition of the atmosphere. No one is sure how much soil has eroded, how much hazardous waste has been dumped, how much groundwater has been polluted.

So are the predictions of *The Limits to Growth* coming true? No, because the book contained no predictions. It showed twelve different futures, all possible we think, some terrible, some terrific. As we saw it then and see it now, the future is not cast in concrete, to be predicted; it is full of potential, to be chosen. *The Limits to Growth* was not about prediction. It was about choice.

Choice is constrained by planetary laws, of course, and the book did say that some choices are physically impossible. Eternal expansion of the human economy is one of them. Trying to cram more and more people, homes, factories, croplands, vehicles, mines, and dumps onto this finite planet will certainly run us into limits.

There are many limits—environmental, economic, social. There is a limit to the amount of oil under the ground and a limit to the amount of carbon dioxide the atmosphere can hold before the planet warms, the seas rise, and the climate changes. There is a limit to the amount of injustice people will tolerate. There is a limit to the ability of human beings to manage complex systems.

No one knows precisely where the limits are or which ones come first. But we do understand something about how they work. That's what *The Limits of Growth* was about. These were our major points:

- One limit may be overcome by conservation, substitution, technical advance, or regulation, but if growth continues, another limit will be encountered—or the same one reencountered. Cutting pollution per tailpipe in half while doubling the number of tailpipes adds up to no progress on air quality. Between 1972 and 1985 Americans spent $70 billion on sewage treatment plants but are emitting nearly the same amount of organic waste to streams because sewage production has grown as fast as sewage treatment.
- If problems induced by limits are solved by sweeping them under the rug, into the water or soil or atmosphere, over to the poor, or off to the future, those problems have not gone "away." They will come up again, somewhere, later, harder, often all at once. High smokestacks transform local pollution into distant acid rain. Sludge that's barged farther out to sea just takes longer to wash back to shore. DDT banned at home but sold abroad comes back in coffee and other imports.
- An economy that depends heavily on nonrenewable resources such as fossil fuels and that degrades renewable resources such as soils and forests is steadily *lowering* its limits to growth.
- There are no clear signals to tell us where we stand relative to global or local limits. The signals are complex, noisy, and delayed. The 1988 drought could have been greenhouse warming; it could also have been just a dry year.
- Even if the signals were clear, we could not act on them quickly. It took sixteen years from the first warning of atmospheric ozone depletion to the first international agreement curtailing ozone-depleting pollutants. It will take ten years to implement the agreement fully. At that point we will begin to know whether we curtailed enough.

In short, it's as if we were driving our economy toward a set of barriers an unknown distance ahead. We can't see clearly, and we can't brake quickly. There is nothing about our price system, technology, forecasting ability, or political wisdom that will guarantee a soft landing. Quite the contrary, they are all encouraging us to accelerate. What we need to do instead, we said in *The Limits to Growth*, is slow down.

That message offends conventional wisdom, but I think it was valid then and is valid now. We wrote it not to predict doom but to challenge the myth of growth as the answer to all problems. We wanted to encourage a search for other answers. How can progress be defined in terms other than expansion by a certain percentage per year? How can human beings find ways of living that are fulfilling, equitable, and consistent with the laws of the planet?

If I would make any change in the book now, it would be to explore further the alternatives to growth—to say more about what a sustainable society could be like. A lot has been learned about that in the past seventeen years.

Tens of thousands of farmers in the United States, Europe, and even the Tropics now get consistently high yields without drenching the countryside with pesticides or chemical fertilizers, and with great reductions in soil erosion. Newly designed appliances, lights, and motors can provide the same services with one-half or one-third as much energy (thereby reducing greenhouse emissions, acid rain, and urban air pollution). In some places, most notably Europe and Japan, recycling municipal waste is becoming a fine art. Industries are learning to reuse hazardous chemicals they once threw out.

Those are all steps that extend limits. They buy time for the longer-term challenge of controlling growth.

There is progress on that front too. Many countries, from Panama to South Korea to Mauritius to China, are successfully reducing their population growth rates. And many people are finding purposes for life more satisfying and less damaging than endless material accumulation.

I think the most remarkable conclusion of *The Limits to Growth* was that the earth could support indefinitely at least 6 billion people, all at a comfortable European standard of living. In seventeen subsequent years of working with global resource statistics, I feel more sure of that conclusion now than I did then. I see no reason why a sustainable world need be impoverished, dull, unjust, technically stagnant, tyrannical, or offensive to the human spirit. Quite the contrary, I think learning to live in harmony with one another and with the limits and cycles of the earth is a far more exciting and worthy goal than mindless swelling.

I don't know if we are running into the limits to growth, or if they are still ahead, or if we have surpassed them and are only now getting the signals that we have gone too far. No one knows. The only certainty is that we are closer to them than we used to be. The global warming, the dying trees, the polluted waters are not a surprise, not bad luck, not minor inconveniences that can be quickly fixed. They are powerful signals about how a finite planet responds to a species that refuses to set its own limits.

That is our choice—to set our own limits rather than have the planet set them for us. A good way to start would be by questioning our vague, sacrosanct goal of growth. Growth of what? For whom? For how long? At what cost? Paid by whom? Paid when?

If we can begin to answer those questions with some specificity and honesty, we may discover what actual desires are unaddressed by the myth of growth. We may begin to define a future we really want. We may discover—I think we will—that there is such a thing as enough.

AMAZING
NUMBERS

SOME PEOPLE love numbers, some people hate them. I love them. I spend my days wallowing in the world's statistics. I learn a lot from numbers. At the same time I'm well aware that statistics are full of pitfalls. They look more precise than they really are. They sweep useful details under aggregate rugs. And numbers simply cannot measure all that's important in the world.

So when I write columns about numbers, which I often do, I am usually pursuing two themes, which at first sound contradictory.

On the one hand, I try to educate people to be more at home with numbers. I wish I could be sure the electorate grasps the difference between a million, a billion, and a trillion. I wish everyone knew what the world's population is and how fast it is growing. Mastering some of the important numbers that chart the state of the world seems to me a minimum requirement for global citizenship.

On the other hand, I often write to debunk numbers and our society's overfascination with them. That's to avoid going too far in the other direction, toward believing only numbers, only measurements, and not the experiential evidence before our very noses. Numbers can be false. Numbers can be misleading. Numbers measure only quantity, when often the real issue before us is one of quality.

One of my favorite sayings, which inspires my column, "The Global Citizen," in many ways, is: "The opposite of a small truth is a lie. The opposite of a great truth is another great truth." In this modern quantitative age, we need to be more numerate *and* less obsessed by numbers, both at the same time.

The World in Two Pages

THE INTERACTION Council is a group of ex-heads-of-state from many countries. Its members include Pierre Trudeau of Canada, Helmut Schmidt of West Germany, and others from Europe, Asia, Africa, and Latin America. The Interaction Council is a forum where these people, who are retired but who still carry considerable clout, can update themselves about weighty matters, debate, and work out possible policies to take back to their home governments.

So I was flattered to be asked in 1985 to prepare a paper for the Interaction Council about the state of the world's resources and environment. There was just one problem. "Be sure not to go over two pages," I was told. "They never read anything over two pages."

Two pages on the state of the planet? Well, I did the best I could. Here's what I came up with—the world's shortest report on the state of resources and the environment (the numbers have been updated to 1990).

Each *day* on this planet 35,000 people die of starvation, 26,000 of them children. This human toll is equivalent to 100 fully loaded 747-jets crashing every day. It is the same number of deaths every three days as were caused by the Hiroshima atomic bomb explosion. And each day, because of population growth, there are 220,000 more mouths to feed.

Yet enough food is *already* raised each year to feed not only the current human population of 5.2 billion, but also the population of 6.1 billion expected by the year 2000.

Each *day* 57 million tons of topsoil are lost to erosion, enough to

37

cover more than my entire town of Plainfield, New Hampshire, to a depth of 8 inches. Each *day* there are 70 square miles more of desert, each four years an area greater than West Germany.

Yet the amount of food produced on the planet has doubled in the past thirty-five years. Hundreds of thousands of farmers know and practice agricultural technologies that preserve the soil, minimize the use of harmful chemicals, and still produce high yields. If their techniques could be widely adopted, world food production could be doubled again.

In the Third World 60 percent of the people do not have access to clean drinking water, which causes billions of preventable illnesses, infections, and deaths each year. One-fourth of the world's freshwater runoff is now made unusable by pollution.

Yet the amount of money that could provide clean water to everyone is only one-third the amount the world spends on cigarettes. The annual stable freshwater runoff of the planet is sufficient to supply double the present rate of human use—more if water is conserved or if unnecessary pollution is stopped.

Each *day* there are 80 square miles less of tropical forest. The annual loss of forest is equal to an area larger than Maine or Indiana. This forest loss results in soil erosion, flooding and drought, siltation of water reservoirs, extinction of species, and enhancement of the greenhouse effect.

Yet much of that deforestation is economically unviable, sustained only by the subsidies of governments that do not understand the direct economic value of a living forest. Saving the forest would actually make money for some of the poorest nations of the world.

Each *day* 60 million barrels of oil—which is nonrenewable—are burned. We pay for it in spills, toxic wastes, foreign debt, urban air pollution, acid rain, and the release of carbon dioxide into the atmosphere at such rates as to threaten a global climate change.

Yet the world could produce all its current goods and services with at most one-fourth of the energy it now uses just by using it more efficiently. Two thousand times our total global energy consumption arrives free from the sun each day; it is infinitely renewable and nonpolluting.

Each *day* on this planet $2 billion is spent on armaments.

Each *day* between ten and one hundred species of life become extinct because their habitats have been destroyed by human activity.

For the first time in history over 50 percent of the human race is literate. Seven full-time television channels and 31,000 simultaneous telephone circuits now link 109 nations through 14 satellites in earth-synchronous orbit. United Nations data systems are providing the world's first standardized, comprehensive information on population, environment, and economic activity. Our ability to gather and communicate information has never been greater.

The world as a whole still has more than enough resources to meet all human needs. Never before has the human population had such power, knowledge, organization, and riches with which to manage those resources wisely and to meet those human needs sustainably. Simultaneously, never before have so many resources been wasted and destroyed on such a large scale in so many parts of the planet or have so many people lived lives of deprivation and suffering.

I never heard whether the Interaction Council did anything with this information. But the important question is, what will the rest of us do with it?

Trivial Pursuit for Global Citizens

S OMEONE ONCE suggested that Americans should have to pass a test, like a driver's license exam, before being issued a passport. The test would measure knowledge of the world—a kind of global citizen's exam.

What should such a test include? The question boggles the mind. Just to make sense of the nightly news, one has to know such things as what is a Shiite, and where is Romania, and through what port does food aid enter Ethiopia. Clearly we need to know a lot to travel wisely in the world.

To prepare ourselves, we must therefore use the most powerful educational technology ever invented: the Trivial Pursuit game. Just think, instead of learning trivia about movie stars and athletes, we could be learning trivia about our planet—*significant* trivia, if you'll pardon the contradiction.

I'd be happy to draw up a contract with the Trivial Pursuit people any time. Here are a few demonstration questions. They focus on the number, distribution, and quality of life of the world's people. If you get even half the answers right, you deserve a gold passport.

QUESTIONS

1. How many people live on the earth?
2. How many more people will be added to the earth's population this year?
3. What percentage of the world's people live in the Third World?
4. Is the population growth rate of the world going up or down?
5. Name the five largest countries in terms of population.
6. Name the five largest countries in terms of land area.
7. Name the five largest countries in terms of economic output.
8. In terms of income per capita, where does the United States rank among the world's nations?
9. What percentage of the world's adults can read and write?
10. What percentage of the world's people are seriously undernourished?

ANSWERS

1. As of mid-1989 there were 5.2 billion people on earth. If you shook hands with one each second, it would take you over 150 years to shake hands with all of them.
2. About 90 million more people were added in 1989, more than the population of Mexico or of ten New York Cities.

3. More than 75 percent of the world's people live in the Third World. That percentage is increasing because 90 percent of the population growth is taking place in the poorest countries.
4. The rate of population growth was nearly constant during the 1980s at about 1.75 percent per year. Before that the growth rate had been going down, from a peak of 2 percent per year in the early 1970s. That decrease of just 0.25 percent makes a difference of 13 million fewer people added in 1989.
5. As of mid-1989 the top five countries in population were:

Country	Population (millions)	% of World Population	% Annual Growth Rate
China	1,104	21.0	1.4
India	835	16.0	2.2
U.S.S.R.	289	5.4	1.0
United States	249	4.6	0.7
Indonesia	185	3.5	2.0

One person in five in the world is Chinese; well over half the world's population is Asian.

6. The five largest countries in land area are:

Country	Area (sq. miles)	% of the World's Land Area
U.S.S.R.	8,599,000	15.0
Canada	3,852,000	6.4
China	3,691,000	6.2
United States	3,615,000	6.0
Brazil	3,286,000	5.5

7. As of 1989 the top countries in total economic output were:

Country	Total GNP (billion $)	% of World Output
United States	4,585	26.0
U.S.S.R.	2,420	14.0
Japan	1,943	11.0
West Germany	889	5.1
France	720	4.1

If you put China on the list, count yourself correct. It is just about impossible to measure China's economy in dollar terms. By some measures China's output ranks higher than West Germany's.

8. According to the World Bank, in 1989 the U.S. average per capita income was $18,430, which gave it second position in per capita income after Switzerland ($21,250) and just ahead of Norway ($17,110), Japan ($15,770), and Sweden ($15,690).

9. About one-half the world's adult population is literate.

10. Estimates of the number of chronically undernourished people range from 200 million to 800 million. If you guessed anywhere from 4 to 15 percent, count yourself correct.

How Much Is a Billion Dollars?

THE SAVINGS and Loan bailout may cost us $500 billion. The 1991 deficit is forecast at $293 billion. The "peace dividend" could be as high as $150 billion a year, if the government were inclined to give it to us, which it isn't.

All these billions of dollars whizzing about in the news are vitally important, I'm sure. As taxpayers and citizens we ought to keep our eyes on them. But personally, I have a lot of trouble thinking about dollars in the billions.

So I decided to make a crib sheet for myself. I keep a list of all the billion-dollar figures I hear. Every time I run across a new one, I add it to the list to see where it fits relative to the cost of a nuclear power plant, say, or a Trident submarine, or Medicare.

The list is turning out to be not only handy, but revealing.

Here it is in three parts: (1) an orientation session on the meaning of

$1 billion, (2) some items that cost only a few billion, and (3) the really Big Bucks.

THE SIZE OF A BILLION DOLLARS

One billion dollars is one thousand million dollars or $1,000,000,000.

A 4-inch stack of dollar bills amounts to $1,000. A billion dollars would be a stack of dollar bills 62.5 miles high. President Reagan used a similar illustration back in 1981. He was talking about a 62-mile-high stack of thousand-dollar bills, which makes $1 trillion. That was before he added the second and third trillion to the national debt.

One billion dollars is the amount the average American worker, with an annual salary of $19,460, will earn, before taxes, in just 51,387 years.

One billion dollars would pay *all* the expenses of my hometown (pop. 2,500)—all the school, garbage, snowplowing, road repair costs—for 750 years.

U.S. military spending disposes of $1 billion in about 32 hours.

BARGAINS FOR ONLY A FEW BILLION

One Trident submarine, which carries enough nuclear warheads to eliminate every major city in the U.S.S.R., costs $1.7 billion. We are building a fleet of twenty-five of them.

The U.S. tobacco industry spends $2 billion for advertising in one year.

President Reagan requested $2.5 billion over five years for his acid rain program (to demonstrate techniques for controlling sulfur and nitrogen oxide emissions from coal plants)

The total 1987 output—gross national product (GNP)—of the economy of Nicaragua was $2.9 billion.

It will take $3 billion to fix the technical problems of the 100 B-1 bombers the U.S. taxpayers purchased for $28 billion.

The 1987 GNP of Ethiopia was $5.9 billion.

The cost of the Seabrook nuclear power plant was $6.5 billion. Estimates of the cost of decommissioning it forty years from now range from $1.2 to $3 billion.

The fundraising target for a five-year program planned by the World Resources Institute, the World Bank, and the U.N. Development Program to preserve the world's tropical forests is $8 billion.

The world's consumers spent $8.7 billion in 1986 buying products of the Coca-Cola Corporation.

The World Bank invests $13 billion each year in economic development projects in the Third World.

The United States budgeted $14.5 billion for foreign aid in 1986, of which $8.5 billion was military assistance and $6 billion was economic assistance. Only $2.1 billion went for economic development projects in truly poor countries.

In 1986 $17 billion was spent to support 4 million American families—including 7 million children—in the Aid to Families with Dependent Children program.

BIG TICKET ITEMS

The total 1986 revenue of the Chrysler Corporation was $22.6 billion, of which $1.4 billion was profit. Chairman Lee Iacocca received $20 million in salary and stock options, a number too trivial to mention in a billion-dollar list, but it keeps sticking in my mind.

The 1987 GNP of Libya was $22.5 billion.

The planned Star Wars budget for 1985–1991 is $33 billion.

The Brazilian foreign debt is $105 billion.

The interest paid by U.S. taxpayers in 1986 on the government's $2,100 billion debt was $136 billion.

The defense budget of the United States for 1986 was $273 billion.

The 1986 government expenditure for Social Security, Medicare, and unemployment insurance was $288 billion.

The 1987 GNP of China was $331 billion.

The 1987 GNP of the U.S.S.R. was $2,420 billion.

The 1987 GNP of the United States was $4,585 billion.

This list suggests many conclusions. You have probably drawn some of your own. My conclusion is that a lot of us little folks, earning our wages, paying our taxes, and buying Coke and Chryslers, generate amazingly huge flows of money. Most of these flows are not being directed toward creating peace or justice or a healthy environment.

If we ever do decide to create these things, it looks as though there will be enough money.

Have You Got a Minute?

I READ SOMEWHERE that 8,500 McDonald's hamburgers are sold every minute.

How on earth could they know that? I wondered. What, if anything, does that number mean?

Of course it can't mean that precisely 8,500 hamburgers—no more, and not one Big Mac less—are sold each and every minute. The actual numbers—if they could be measured—must look more like 8,532 and 8,489 each minute. They probably vary more widely than that—surely hamburgers are more popular at 6 P.M. than at 6 A.M. Though McDonald's may be so distributed around the world that the sun never sets on the Golden Arches, still they couldn't cover the time zones evenly enough to balance out all the diurnal peaks and troughs.

What that 8,500 must mean is that some bright statistician at the home office took the monthly hamburger sales figure, divided it by 31 (or 30 or 28), and then by 24 and then by 60 and came up with 8,500. A more accurate but far less interesting way of communicating the same information would be, "If you take the reported average monthly sales figure for McDonald's hamburgers, which is roughly 370 million, and divide by the number of minutes in a thirty-day month, you come up with an average per minute hamburger sales rate of 8,500, rounded off to the nearest hundred."

Why express hamburger sales in per minute terms? Because numbers like 370 million are unimaginable. They make people's eyes glaze over. I can't begin to picture 370 million hamburgers. I can hardly handle 8,500 per minute, but at least that number punches into my consciousness the intended message, which must be something like: "McDonald's puts out a humongous lot of hamburgers, day in, day out, minute in, minute out. What a big and important and productive company it must be."

Enormous numbers do become more digestible when expressed in per minute terms. Take the government deficit, for instance. That number has been running roughly $200 billion per year, a number that apparently puts the electorate to sleep, since we haven't yet thrown the rascals out. But—let me get out my calculator here—$200,000,000,000 divided by 365 is $550 million per day, divided by 24 is $23 million per hour, divided by 60 is $380,000 per minute. Every minute, day and night, your government is borrowing $380,000. Does that make it any easier to comprehend the hole that is being dug for you and your future taxes?

Anyone can play this per minute game, and many do. The Hunger Project publicizes widely the "fact" that twenty-four people starve to death every minute, eighteen of whom are children.

How does it know that? It starts with United Nations' annual worldwide mortality statistics, which are pretty uncertain, especially in the poorest parts of the world where hunger is greatest and statisticians rarest. (I was told once that Burma determines its infant mortality rate by looking up Thailand's and adding 50%.)

Mortality rates may be uncertain, but the United Nations does the best it can to keep them roughly in the ballpark, and they are slowly improving in accuracy. In case you're interested, the numbers say that roughly 50 million people died from all causes in the world last year. That's ninety-six per minute.

Now how much of that mortality is due to hunger? A wild guess is required here—"hunger" is almost never given as a cause of death on a death certificate. Hungry children die of measles and diarrhea and influenza that would not have killed them if they had been properly nourished. Most of them die without benefit of a death certificate anyway. U.N. experts in field stations all over the world estimate what fraction of deaths could have been prevented if malnutrition had not been a factor. Their figures range from 13 to 17 million deaths each year because of hunger-related causes.

The lowest figure, 13 million, comes to 24.7 deaths per minute. The Hunger Project's figure of twenty-four per minute is a conservative one. If expressing that horrendous number in per minute terms makes it any more real to people, more power to the Hunger Project.

Some big numbers are thrown around about the loss of tropical forests. In the early 1980s those numbers were so wildly different, source by source, that nobody knew what to believe. Then the Food and Agriculture Organization (FAO) of the United Nations took on a global study using satellite mapping, aerial photography, and ground surveys.

It came out with an official figure of about 18 million acres of tropical forest lost each year. That's 34 acres a minute, an area the size of my farm every two minutes, an area the size of Maine or Indiana every year. That number—18 million acres—is, by the way, related to the sales rate of hamburgers. In Latin America a major reason for forest clearing is to graze cattle to provide beef exports for gringo fast-food chains. And in the last two years, the FAO says, the number has risen greatly.

We could go on. Every minute 162 people are added to the world's population. Every minute 60 million barrels of oil are burned. Every minute $1.5 million is spent on armaments.

It took you somewhere between two and five minutes to read this column.

The Ten Best Nations in the World

SUPPOSE THAT each year there were an award for the world's best nation. If you were one of the judges, how would you make your decision? Would you rank nations by their wealth, their military power, or their civil liberties? By lawfulness? By cleanliness of the environment?

The question is not trivial. People and governments do, consciously or unconsciously, rank nations on many different scales. Those rankings have an enormous influence on decisions about how to spend government money, about what laws and regulations are needed, about where to direct human energy.

The most widely accepted index of a nation's achievement is its gross national product (GNP) per capita. Governments quote it, pride themselves on it, do all they can to make it go up, and look embarrassed when it goes down. But few people are really quite sure what GNP per capita means.

The GNP is the total amount of money spent in a year by the country's consumers on all goods and services. The GNP *per capita* is the GNP divided by the population; it is the money value of the stuff that passes through an average person's life in a year—"stuff" includes food, shoes, cars, dentist bills, legal fees, and that person's share of bureaucrats' salaries and the defense budget.

Per capita GNP is roughly related to human welfare, but only roughly. It counts only those aspects of life that can be measured in money terms and only consumption that is cash-mediated. If you grow vegetables at home instead of buying them in a store, for instance, the GNP goes down, though the quality of your life may go up. GNP includes some bads as well as goods. For example, automobile accidents increase GNP because they increase hospital expenses, repair bills, car purchases, and insurance premiums. And per capita GNP says nothing about the equity of the distribution of goods over the population.

To illustrate the strengths and weaknesses of GNP per capita as a measure of a nation's quality, here are the ten top winners in the best nation competition for 1985, as measured by GNP per capita:

Country	GNP per Capita
1. United Arab Emirates	$21,340
2. Qatar	$21,170
3. Brunei	$21,140
4. Kuwait	$18,180
5. Switzerland	$16,390
6. United States	$14,090
7. Norway	$13,820
8. Sweden	$12,400
9. Saudi Arabia	$12,180
10. Canada	$12,000

The four Middle Eastern countries that top this list produce a lot of oil and spend a lot of money each year, but that money does not necessarily reach all citizens or buy a high quality of life. And another problem with GNP as a measure: all the Arab countries on this top ten list had dropped off the list by 1990, simply because the price of oil went down in the late 1980s. Per capita GNP is a much more volatile figure than is actual quality of life.

Some international organizations, including UNICEF and the Hunger Project, rank nations by a very different index—the infant mortality rate (IMR). The IMR expresses how many babies, out of every 1,000 born, die before they reach their first birthday.

The United Arab Emirates, first on the GNP per capita list, has a mediocre IMR of 45. That means 45 babies out of 1,000 die during the first year of life. By comparison, Saudi Arabia has an IMR of 103—a loss of one baby in ten. The IMR of the United States is 10.5—one baby in 100 does not survive to age one. The world's highest IMRs are in Afghanistan, Sierra Leone, and the Gambia—all around 200.

The infant mortality rate is perhaps the most sensitive single measure of a society's quality of life. It reflects the general level of nutrition and health care. It is affected by water quality, the quality of housing, and the level of education, especially the education of women. It shows the nation's concern for its most ordinary and helpless citizens, its babies.

A high GNP per capita is usually associated with a low IMR, but not always. Libya, with a GNP per capita of $7,500, has a shamefully high IMR of 92. Conversely, China has one of the lowest per capita GNPs in the world—$290—but also an amazingly low IMR—38. That puts China in the same league with the much richer Soviet Union, which has a GNP per capita of $6,350 and an IMR of 32.

If we awarded the 1985 ten best nations prize on the basis of IMR, only four of the GNP per capita winners would make the list—and the United States would not be among them:

Country	IMR
1. Finland	6.0
2. Japan	6.2
3. Sweden	7.0
4. Iceland	7.5

Country	IMR
5. Switzerland	7.7
6. Norway	7.8
7. Denmark	8.2
8. Taiwan	8.9
9. France	9.0
10. Canada	9.1

All these countries are relatively rich in material consumption, though material consumption is not necessarily their highest priority. All invest heavily in human services and distribute those services fairly equitably. All of them have low birth rates—the kind of society that cares for its babies does not seem to generate runaway population growth. All have stable governments and relatively high levels of democracy and civil rights. This may be as acceptable a list of the ten best countries, at least by my set of values, as any single indicator could produce.

If the nations of the world competed not to build the biggest weapons or the biggest gross national products, but to take the best possible care of their babies, it would be a very different world indeed—and a better one.

We Measure What We Care About

WHEN THE 1989 World Bank Development Report came out, the event created excitement only in the hearts of statistics freaks like me. No sooner had my copy arrived than I turned to my favorite table, Table 30, Income Distribution. I like to keep up with how the top 10 percent lives.

To my disappointment the numbers were exactly the same as 1988's, not because the world hasn't changed, but because the World Bank, once again, hasn't updated that table. The numbers in Table 30 are fascinating, but many of them are now at least ten years old.

That in itself is worth noting. The world's keepers of economic statistics chart with eager concern the total amounts of money that flow around the world, the rate at which those flows increase, and what countries owe how much to whom. But they seem uninterested in who ultimately gets it, in how equitably income is distributed on its way to what is presumably its purpose—the enhancement of human welfare. (That *is* its purpose, isn't it?)

I don't see how they can be uninterested. Even with old numbers, even with only forty-six out of 120 countries reporting, Table 30 contains dynamite information.

It says, for example, that in the United States the richest 10 percent of households spend more money than the poorest 40 percent all together—the top 10 percent get a 23.3 percent share of national income; the bottom 40 percent get 17.2 percent. Those are 1980 statistics The distribution has swung further toward the rich since then, but the World Bank hasn't noted that yet.

In Japan in the latest year listed, 1979, the richest 10 percent got almost exactly as much of the pie as the poorest 40 percent: 22.4 percent to 21.9 percent, respectively.

If you think that's unfair, consider Zambia (in 1976), where the richest 10 percent got 46 percent of the income stream and the poorest 40 percent got only 11 percent. Then there's Brazil, the world inequity champion (among the countries for which there are statistics). In Brazil in 1972 (why no update since then?), the richest 10 percent garnered 51 percent of the national income. The poorest 40 percent got by somehow on just 7 percent.

UNICEF also puts out an annual report, called *The State of the World's Children*. It focuses firmly on the welfare of people, especially young people, who in every country live disproportionately in the bottom 40 percent of the barrel. UNICEF suggests that countries should be ranked not by average GNP per capita, but by the average income of the lowest 40 percent. According to the World Bank's Table 30, that would change national rankings considerably.

Kenya, for example, has an average GNP per capita double that of Bangladesh. But the poorest 40 percent of Kenya's people get only 8.9

percent of the national income, whereas the poorest 40 percent in Bangladesh get a 17.3 percent share. That means there is no difference in the actual income levels of the poor in the two countries—they are both at about $65 per household per year, the lowest of all the reporting countries.

Mexico and Hungary fall close together in a GNP per capita ranking, each at about $2,000 per person. But the average income of Mexico's poorest 40 percent is only about half that of Hungary's ($515 and $994, respectively).

The United States ranks higher than Japan in average income per person ($18,530 to $15,760), but lower in the average income of the bottom 40 percent. (Japanese families in the bottom 40 percent average $9,022, U.S. families only $7,967.)

All these numbers have to do with distribution of income within nations. What if we looked at the income distribution *across* nations? How is the pie shared over the whole world? What proportion do the *world's* upper 10 percent get?

We have no way of knowing. Only forty-six of 120 nations are reporting. The World Bank does not tabulate income distribution worldwide.

Why not? Why do the world's statisticians know so much about total money flows and so little about money distribution? Why is Table 30 never complete or current?

The only explanation I can come up with is that the people in the world's upper 10 percent (including every economist at the World Bank) really don't want to know.

Nothing Is So Powerful as an Exponential Whose Time Has Come

THE REASON environmentalists are so often gloomy is that they know what the word "exponential" means.

"A lack of appreciation for what exponential increase really means leads society to be disastrously sluggish in acting on critical issues," said Dr. Thomas Lovejoy of the Smithsonian Institution in a speech that has been reverberating through the environmental community. "I am utterly convinced that most of the great environmental struggles will be either won or lost in the 1990s, and that by the next century it will be too late."

What is he talking about? What does "exponential increase" mean?

It means growing like this: 1, 2, 4, 8, 16, 32. Doubling and then doubling again and then doubling again. Everyone understands that, right?

Not really, not at a gut level. For example, suppose you agree to eat one peanut on the first day of the month, two peanuts on the second, four peanuts on the third, eight peanuts on the fourth, and keep doubling every day. How long do you think you can keep going? How long will a pound can of shelled peanuts last you?

The first pound of peanuts will be gone on the ninth day; you'll eat half the can that day and feel pretty queasy. On the tenth you'll eat a whole pound, if you can, which I doubt. By the fifteenth you'll be scheduled to eat 32 pounds of peanuts. You'll have to eat roughly your own weight in peanuts by the seventeenth; on the twenty-first day the total will have risen to one ton; and by the end of the month, assuming a thirty-day month, it will be 500 tons.

Just a few doublings add up ferociously fast—that's what Lovejoy was saying.

Mexico, with a population of 84 million and a doubling time of

Figure 1

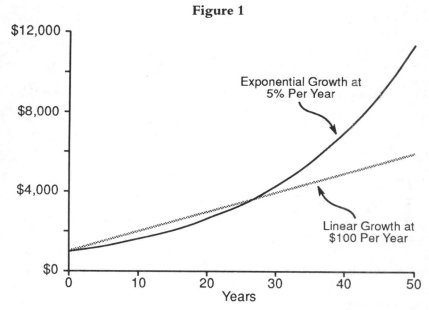

A thousand dollars to which 100 dollars is added each year grows linearly. A thousand dollars to which 5 percent is added each year grows exponentially. At first the linear growth goes faster, but the exponential catches up and surpasses it.

twenty-nine years, will, if it keeps that up, grow to 168 million in twenty-nine years and to 672 million within the lifetime of a child born today. That's nothing compared with Kenya, which has a doubling time of seventeen years. If it keeps growing at that rate, in seventy years there will be ten Kenyans for every one today.

Until the 1970s world oil consumption was growing at 7 percent per year. That means doubling every ten years. (The doubling time of anything growing exponentially is 70 divided by its annual growth rate—70 divided by 7 percent is a ten-year doubling time.) Every ten years we used as much oil as we had used in all previous history. Every ten years we had to go out and discover as much oil as we had ever discovered before—and then, to keep going, discover twice that much in the next ten years.

We didn't keep going. We couldn't have. Exponential growth makes the cupboard bare very fast. Even if the entire earth were filled with nothing but high-grade crude oil, if we used it with an annual growth rate of 7 percent, it would be gone in 342 years. There's still plenty of oil around now, but we've been burning it faster than we've been discovering it for twenty years.

You may have heard that we have 1,000 years' worth of coal. If we burn 7 percent more of it each year than the year before (which we may well do, substituting it for the disappearing oil), it will last just sixty-one years, and it will bring on a global climate change much faster than even the worst pessimists are now expecting.

Said Lovejoy in 1988, "I find to my personal horror that I have not been immune to naivete about exponential functions. While I have been aware that the . . . loss of biological diversity, tropical deforestation, forest dieback in the northern hemisphere, and climate change are growing exponentially, it is only this very year that I think I have truly internalized how rapid their accelerating threat really is."

You don't get much reaction time when your problems grow exponentially. My favorite story to illustrate that point is an old French riddle.

Suppose you own a pond on which a water lily is growing. The lily doubles in size each day. If the lily were allowed to grow unchecked, it would completely cover the pond in thirty days, choking off other forms of life in the water. For a long time the plant is almost invisible, and so you decide not to worry about cutting it back until it covers half the pond. On what day will that be?

On the twenty-ninth day.

We are emitting carbon dioxide and several other greenhouse gases into the atmosphere exponentially. We are clearing tropical forest at an exponential rate. The human population is growing exponentially. Human energy use, human production of synthetic chemicals, deserts, and trash are growing exponentially. Our economy is growing exponentially, and we cheer it on, although an economic growth rate of, say, 3.5 percent per year means another whole industrial world plopped down on top of this one in just two decades.

We can't keep it up. If we understood the consequences of exponential growth, we wouldn't even want to try.

POPULATION
AND ABORTION

THE MOST important process going on in the world is the hardest one to write about—population growth. It is continuous, gradual, exponential, and overwhelming. If it is not contained, it will undermine all environmental and economic progress. I am convinced that it can be contained but only if we understand and work at it with seriousness and compassion.

In the world of journalism, population growth is *boring*. Writing about it is like writing about the movement of a glacier. The message is the same, day after day, year after year. But it's such a critical message that I keep trying.

Another process that is hard to write about, for different reasons, is abortion. That subject is so loaded that every column I write on it brings, no matter what I say, a flood of letters and phone calls, most of them raving. I know of no other subject that stirs up such public irrationality.

Abortion is a minor issue compared with population growth, but the two are snarled together in the American mind. That has led to the tragic interruption of many good family planning programs on the grounds of preventing abortion. So again I keep trying, in this case to undo the snarls.

My personal position on abortion is close to neutral, so I get shot at from both sides. I think abortion is violence; I abhor violence; I have never had an abortion; I would not have had one except under the most

57

extreme circumstances. I would never presume to tell someone else whether to have an abortion or not; I would certainly not threaten or punish a woman for having one. And I am as repelled by the intolerance and hatred with which the abortion debate is carried on as I am by the violence of abortion. Over the years I have become less concerned whether any particular view of abortion prevails and more concerned with the reasons why our society cannot come to peace on this subject and why we let it divert us from the far more important subject of population.

Population and Depopulation—Two Worlds in One

FOR THE first time in history the Census Bureau has projected a slow decline for the U.S. population. It will not happen until well into the next century, after the momentum of the baby boom carries us from our present 248 million people to 300 million. But if our fertility rate stays as low as it is now, even with steady immigration we are headed for a population decline.

We are not the only nation in this demographic position. In seven European nations population growth rates are already zero or negative. Over the next thirty-five years West Germany's population is expected to decrease by 10 percent, from 60 million to 54 million. Sweden's population is expected to go down over that period by 6 percent, Switzerland's by 8 percent.

Yet the world population as a whole is expected to double over the next forty years. It grew by 90 million people in 1989, equivalent to the total population of Mexico, the largest one-year increment ever.

"Every human society is faced with not one population problem but two," said Margaret Mead—"how to beget and rear enough children, and how not to beget and rear too many."

We have both problems in different places, two demographic worlds, one declining, one soaring. They are directly correlated with two economic worlds, one rich, one poor. As they say, the rich get richer and the poor get children. Of the 90 million new human souls added in 1989, 16.3 million were African, 9.6 million were Latin American, a whopping 51.7 million were Asian. Over 90 percent of the world's population growth is taking place in the Third World.

For a population to maintain its numbers, its women must bear on

Figure 2

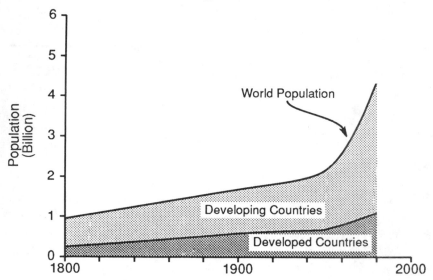

Since 1800 world population has grown from one to five billion. Most of that growth has taken place—and is still taking place—in the developing countries.

SOURCE: World Bank

average slightly more than two children each. In the United States the average number of children born to each woman is 1.8. In Canada it is 1.7. Denmark averages only 1.4 children per woman, West Germany 1.3.

In Mexico average fertility is 4.0 children per woman. In Nigeria it's 6.6, in Kenya 8.0.

The Third World has *added* since 1960 more people than the *total* populations of North America, Europe, the Soviet Union, Japan, and Oceania combined.

About 665 million *additional* jobs will be needed in the developing countries over the next twenty years, just for children who are already born. That is greater than the *total* number now employed in all of North America, Europe, the Soviet Union, Japan, and Oceania.

Since 1950 Nigeria, the most populous country in Africa, has grown from 43 million to 105 million people. In the next thirty-five years Nigeria is expected to add another 206 million people. Barring disaster, that means the Nigerian population will grow from 43 million to 312 million in just seventy years, one human lifetime. The implications for the land, the cities, the economy, the politics, the people of Nigeria and of West Africa—and of the world—are staggering.

India, with one of the longest established family planning programs in the world, has a slow population growth rate, for Asia. Its women average 4.3 children each. If, as the United Nations's forecast assumes, India's birthrate continues to come down, its population will grow from 735 million to 1,200 million over the next thirty-five years. That will amount to a population nearly five times that of the 1989 United States on one-third the land area.

The numbers would be even more sobering if the world had not made tremendous progress over the past twenty years in economic development and family planning. Average fertility in the Third World has dropped from six children per woman to four. Contraceptive use has increased from 9 percent of married women of reproductive age to 43 percent. The only places where birthrates are not dropping are the very poorest places on the planet.

Why is rapid population growth so devastatingly correlated with poverty?

One theory has it that people are poor because they go on reproducing and dividing their land, their food, their everything over too many children. Population growth makes poverty. Another theory reverses the causation: Poverty makes population growth. Poor people have many children because children are needed to work and to support their elders and because children don't cost much if you don't have to buy them Reeboks and send them to college. Having children is one of the few powers the poor can exert over their own lives, and one of the few hopes of getting ahead.

There is a third theory. World fertility surveys indicate that anywhere from a third to a half of the babies born in the Third World would not be, if their mothers had access to cheap, reliable family planning and had the personal empowerment to stand up to their husbands and relatives and choose their own family size. Economic development brings lower birthrates because it brings to women the pill, literacy, and self-determination.

All three theories are probably right. Poverty plus unempowerment plus population growth make a consistent set and a formidable trap. The only way out is economic and personal advancement. The rich one-fifth of the world is living testimony to the fact that some mixture of opportunity, health, and family planning does bring population growth rates down. When non-European people such as the Japanese, Singaporeans, and Taiwanese experience economic development, their fertility also goes down (to 1.8, 1.7, and 2.1 children per woman, respectively).

Social services, especially to women, not wealth per se, seem to be the key to lower birthrates. The Chinese, though they are one of the poorest peoples of the world, have brought their fertility down to 2.4 children per woman, partly by social coercion, but mostly by broadly available education, health care, and family planning. Saudi Arabia and the United Arab Emirates, on the other hand, two of the world's richest countries and most oppressive to women, average 7.1 and 5.9 children per woman, respectively.

Can the trend to lower birthrates become too much of a good thing? Is the population decline of the developed world something to worry about?

Most people in countries with low fertility are taking the prospect calmly. Smaller populations in the developed countries will help with many problems, from unemployment to solid waste to acid rain. In crowded, polluted Europe and in the United States it's hard to argue that more people are needed. Peter von Ehr, a German sociologist, says, "If we want to keep our present standard of living, the last thing we need is more children."

As for maintaining a Caucasian presence in global affairs, demographer Lincoln Day remarks, "A billion or so Europeans cannot help but figure prominently on the world stage."

The one possible objection (aside from racist ones) to a slowly declining population is that it will mean relatively fewer young people and more old people. In Sweden 18 percent of the population is under the age of fifteen and 17 percent is over the age of sixty-five. In Kenya 53 percent is under fifteen, 2 percent over sixty-five.

Shifting age ratios should pose no economic burden—society will have to spend more on older people but less on young people. There are problems of adaptation, however. Fewer first-grade classrooms will be needed and more Meals on Wheels, fewer pediatricians and more gerontologists. Above all, declining populations will require a revaluing of human resources of all ages. Lincoln Day says, "More will need to be done to ensure that older people are permitted a life of dignity and reasonable comfort, that society is enabled to take advantage of what older people have to offer, and that the relatively scarce resource of children and young adults is not wasted."

Whatever problems are posed by that revaluing, they will be nothing compared with the problems where birthrates are still high. Kenya is facing a tripling of population over the next thirty-five years. That increase is a result of impoverishment, and it is likely to be a cause of further impoverishment unless something is done to stop it.

What can be done? Anything that will help provide basic needs, equal opportunity, and family planning technology to every person on earth will help. That can include Third World debt relief, fair trade, a foreign aid program truly aimed at the poorest of the poor, full financing of U.N development and population programs, and real support for Third World self-determination. Any action will help that comes out of true concern for the welfare of poor people and that is based not on condescension but on partnership with them.

An old Chinese proverb says, "If we don't change our direction, we'll end up where we're headed." Where we're headed is toward another doubling of world population, nearly all of it in poor countries. We're headed for a greenhouse climate change, for desertification and deforestation, for a world ever more desperate and turbulent. The demographic consequences of our present divided, unequal, and unjust world are clear. We don't have to end up there. We do have to change our direction.

The Thinning of Turnips and the Right to Life

Twenty-fifth of July, plant turnips, wet or dry," the old-time gardeners say. So on that day I plant the fall turnip crop. It's a big planting. I can't get anybody in the house to eat turnips, but I store them in the root cellar and feed them to the sheep all winter long.

The twenty-fifth of July was dry and hot this year. I was afraid the seeds wouldn't germinate in the heat, so I sowed them extra thick and watered them down.

Every last seed came up. I could swear that more came up than I planted. I had a big job getting those turnips thinned.

Thinning is the only garden chore I dislike. Those thriving baby plants, green and promising, are all trying to fulfill their genetic potential, and I must pull up three-fourths of them and leave them to die in the August sun. I have to grab hold of some hard thing inside myself before I can grab hold of the young turnips. I feel like an imposter god, arbitrarily dealing out life and death.

As I worked down the row this year, I was reminded of the protests of the right-to-life people against the children's television program "Sesame Street." "Sesame Street" had done a segment about thinning marigolds, explaining carefully that plants shouldn't be cramped. To grow properly, they need nutrients and water, light and space, "just like children."

"Why thin the marigolds and throw them away?" fumed the Pro-Lifers, reading much more into the message than the kids did. "Why not *transplant* them?"

I smiled at that as I yanked out turnips. If I transplanted all these seedlings, they'd cover my whole farm. It would take months to do the transplanting, and who could use so many turnips?

Of course the protesters were talking about human beings, not marigolds or turnips, and about abortion, not thinning. They would say that practical calculations are fine for turnips, but not for human beings, who are special and precious, to be treasured and nurtured, every last potential one of them.

I sympathize with both sides of this controversy. I also think that each side, untempered by the other, is dangerous.

The Pro-Lifers seem to think of life as a scarce commodity that needs to be protected. But on this planet life is in exuberant abundance. Especially in abundance are seed and young. I don't have to lift my eyes far beyond the turnip row to contemplate the sprouting crabgrass, the potato bug larvae, the fat lambs in my pasture. They are all capable of taking over the farm and ruining its harmony and productivity, if permitted to reproduce without restraint.

The multiplying potential of every species, including Homo sapiens, is immense. Yet one of the great principles of nature is "enough." Only so many turnips can fit in the row and still be healthy, fulfilled turnips. Only so many sheep can graze on the pasture. Whenever I have failed because of softheartedness or busyness to thin the turnips or cull the sheep herd, the result has been spindly turnips, sickly lambs, and an eroded pasture.

Human beings are as subject as turnips to the physical laws of the planet. We can overpopulate and destroy our resource base. Our current numbers are without precedent and growing rapidly. But I get uneasy when my environmentalist friends compare population growth to a locust plague or a cancer spreading over the face of the earth. I understand their point, but I cringe at their analogies.

The equation of populations with plagues can lead well-meaning people to suggest solutions to the population problem that come uncomfortably close to playing imposter gods. Most environmentalists I know are such softies that, like me, they have trouble pulling up a baby turnip, but when they get theoretical and worked up, they can exude an appalling lack of respect for human beings.

It could be simultaneously true that human beings are both part of nature and something special in nature. Our numbers, like those of any other species, must be limited. But we are the only species with the intellect to realize that and to control our numbers voluntarily in a way that respects both the preciousness of each individual and the constraints of the environment.

We haven't yet figured out how to do that. But we have a chance to pull it off, as long as both the idea of "enough" and the idea of the "sanctity of life" are vigorously represented in our society. Neither should be permitted to overpower or silence the other.

Right-to-Life, Right-to-Sex, and Right-to-Family-Planning

THIS YEAR the world's population will increase by about 90 million people. Ninety percent of that increase will take place in poor countries.

Many Third World countries are worried about providing land, education, and jobs for these rapidly growing populations. Many Third World women are now aware that they do not need to bear many children, close together, at the risk of their own and their babies' health. Governments are asking for assistance for population programs; people are asking for family planning. And just as the requests are increasing, the United States is restricting its population assistance programs.

For twenty-five years we have been the most generous source of family planning aid in the world. But in the 1980s our government, in the name of abolishing abortion, cut back sharply programs in population education and contraception. For example:

• Direct use of U.S. aid money for abortion has been illegal since 1973. The Reagan administration extended that prohibition to stop funding any organization that provides abortion services *even with non-U.S. money.* The first victim was the International Planned Parenthood Federation, which lost $17 million per year in U.S. funds for contraception because eleven of its 120 national programs offer abortion services, paid for by other donors.

- Our government canceled funding for the U.N. Fund for Population Activities because, among many other activities, the UNFPA works in China, where abortion is legal.
- Total U.S. appropriations for population aid decreased by $40 million for 1986. The only program that increased was "natural family planning," which advocates the rhythm method, the most ineffective form of birth control.

Watching this damage to programs I consider vital, I can't help but wonder what the Right-to-Life people are really up to.

They can't simply be opposed to abortion. If they were, they would welcome the contraceptive programs they are cutting, all of which demonstrably reduce abortion rates.

They can't be outraged at the deaths of innocent children, or they would be clamoring to support programs for food, clean water, and vaccinations, the lack of which kill far more babies each year than do abortions.

What do Right-to-Lifers really stand for?

I got an insight into that question once when I was moderating a debate between Pro-Life and Pro-Choice activists. After the session was over, I asked each side what it thought the other side *really* cared about. The answers were unhesitating, and they were remarkably symmetrical.

The Pro-Choice people said about the Pro-Lifers: "They are just plain prudes. They want to control everyone else's sex life, to stop the liberation of women, and to punish the act of sex with the burden of pregnancy." The Pro-Life people said of Pro-Choice: "They are secretly ashamed of their promiscuity. They want society to condone their immoral behavior. They do not want to be held to account."

Now there is an interesting discussion here, but it is not the one the two sides had just publicly engaged in. Is the real disagreement about American sexual morality? If so, how did it ever get so twisted as to interfere with foreign family planning programs?

If the Right-to-Life debate is really a Right-to-Sex debate, then let's argue that topic straight out. Let's talk about teenage pregnancies, divorce and broken homes, venereal disease, and the many other social ills arising from irresponsible sexuality. But let's not confuse that subject with international population programs. When there is so much at stake in the Third World, so much poverty and famine, so much hope and resolve, we should not permit family planning programs to be frustrated by our inability to talk out loud about what's really on our minds.

Maybe We Could Agree About Abortion

I HESITATE EVEN to write the word. It creates nothing but trouble. Minds are so rigidly made up, emotional fuses so short, that any mention of abortion touches off nothing but preprogrammed diatribes.

But I feel I have to say something, not about my own position, but about the issue itself, which is fast becoming a national disaster. It causes people who regard themselves as decent and law-abiding citizens to throw bombs, shoot at Supreme Court justices, and yell terrible accusations at other law-abiding citizens. Each side dehumanizes the other, twists facts, uses inflammatory language, and ultimately incites violence.

I am tired of hearing my friends call each other murderers or meddling religious zealots. I would like to see my community and my nation end this energy-draining polarization. I'd like to start the discussion all over again, beginning not with the question, "Should abortion be legal?" but with the question, "Is there anything about abortion that we can agree on?"

I submit that we might agree on enough to work out a set of policies we can all live with.

For example, one point of agreement is certainly, *abortion is awful*. No one has an abortion for the fun of it. No one wants to have one, ever, though some feel they must. Here is common ground. Let us do all we can to minimize the probability that any woman will ever feel the need for abortion.

How could we do that? Here are some suggestions. Much could be done to reduce the occurrence of rape, ranging from better policing to reducing the violent sexual images that permeate our media. We could increase the likelihood that a woman who considers becoming pregnant will get a thorough physical exam to be sure that childbearing will not endanger her health. We could ensure that all women, from the time they functionally become women, have access to the best possible technologies for controlling their fertility. And if all prospective mothers had assurance of education, jobs, health care, and financial security, there would be far fewer who would face the bleak choice between abortion and raising a child in poverty.

Another point of agreement, I expect, is, *life is precious*. Mothers, fetuses, babies, and children are especially precious because they represent future life, the continuity of our species and our civilization. Perhaps we could find consensus around policies that treat mothers and children as if they really were precious.

We could guarantee their basic nutritional needs, especially during pregnancy and lactation. For fifty years Sweden has had a program designed to be sure that no mother or child will lack any material or educational resource. The program includes everything from sex education to free prenatal and postnatal medical care to school lunch programs. Sweden's infant mortality rate is half of ours. So is its abortion rate, though abortion in Sweden is both legal and free.

There is a third possible area of consensus. *People should be free to make their own choices*, especially about childbearing. We all resist outside meddling in decisions about whether and when we marry, use contraceptives, have a child. These decisions do affect the future of the state, however, and of other individuals. We might find agreement that the state's interest is primarily in being sure that everyone has the most complete, straightforward information possible, to make childbearing decisions deliberately and wisely.

In this arena we have a long way to go. If I am a typical example from the previous generation and my students are examples of the current one, most of us blunder into courtship, sex, marriage, and childbearing through a messy confusion of naive hope, pop-tune inanities, parental dicta, and a burning need to be accepted and loved. Rarely is the decision to conceive a child based on even a partial understanding of the responsibilities of bringing up that child.

Our young people should know—and know early—about human reproduction, the development of the embryo and fetus, the nutritional needs of mothers and babies, contraceptive techniques, how babies grow and learn, and what parenting really means. They also need to understand and resist all the messages their society throws at them—principally in advertising—that urge gratification instead of responsibility. They should have some actual experience of caring for a baby before they make one of their own. Programs in some innovative schools indicate that this information is delivered most effectively not by sanctimonious teachers but by older teenagers trained to the task.

What if the question were not, "Shall abortion be legal?" but "How can we ensure that each child is loved, nourished, and provided for from the moment of conception and throughout childhood?" Or, "How can we give each family the maximum possible information and support for the wonderful, difficult job of raising the next generation?"

If we could come together around crucial questions such as these, maybe we could forget our enmity. Maybe we could even put together a society in which the question of abortion is not really important because enough other needs are met that the need for abortion almost never arises.

What Are We Really Fighting About?

TWENTY YEARS ago I was legally forbidden, as a married woman living in Massachusetts, to have an abortion or to use contraceptives. I was a low-income but privileged member of society (a graduate student), however, so I obtained the pill with no trouble. My university health service dispensed it quietly, as did many doctors throughout the state.

I was not personally inconvenienced by the laws (which were over-thrown a few years later by the U.S. Supreme Court), but I considered them unjust and intrusive. So every year I attended hearings on bills that tried—never successfully—to take reproductive decisions out of the hands of the state and to put them into the hands of the family.

Those hearings were always a circus. Picketers on both sides marched in front of the statehouse. Busloads of nuns occupied the front rows of the hearing room, saying their rosaries. Pickled fetuses in bottles were passed around, as were pictures of well-loved babies who would never have been born if. . . . Weeping women, married and single, of all ages, talked about the devastation an unplanned child or a botched illegal abortion had caused in their lives. The one time I tried to testify, the presiding legislators, all male, thought it humorous to ask me prurient questions about my sex life. They did that to all the young women.

The 1989 Supreme Court decision *Webster v. Reproductive Health Services* has guaranteed that such hearings will take place annually in all fifty state capitols. I don't know anyone who is happy about that, but I don't know anyone on either side of the abortion issue who will let it rest. We seem condemned to go on and on defying, accusing, picketing, litigating, fire-bombing, hating, all the while talking of morality.

Why? Why has this particular matter got us so monumentally stuck in hopeless contention?

I suspect that the abortion issue remains unsolvable because the questions we think we are fighting about are not what we are really fighting about. The real questions are larger than abortion, larger than—but connected to—women's rights, prayer in schools, sex education, gun control, affirmative action, all those battle-torn beachheads upon which we are fighting a larger war.

What is that war about? Here's my guess, derived from looking inside to see what it's about for me.

I feel that I am swept up in huge, historic changes, way beyond my control, changes in who we think men and women are, how we believe we should relate to one another, how we should bear and bring up children.

What I learned as a child about the stability of marriage, my role as a woman, and the importance of the family has been deeply shaken. My society pours into one ear prudish standards of behavior and into the other ear hundreds of sexual come-ons in daily advertising. Sometimes I hear that I am liberated; sometimes I am treated as a body with no mind or soul. I am told I have to be virtuous but attractive to men, to bear children but earn a living, to be independent but subservient.

My confusion comes, I think, from Pandora's boxes that were opened long before I was born—three boxes, to be exact: health care, industrialization, and the political idea called democracy. All three brought wonderful benefits and unexpected side effects to which we (I, anyway) still haven't adjusted.

Because of modern health care, spouses live together for a longer time, over more life changes, putting great strains on the idea of marriage-for-a-lifetime. Nearly all children survive, making it necessary to match effective death control with effective birth control.

Industrialization is systematically moving traditional family functions into the commercial market. Bread, garments, repairs, cleaning, even child care and elderly care are now products of industry. Men and women must be workers outside the home, to afford what they used to produce for themselves, which they no longer have time to do, because they have to work outside.

Our democracy began with the words "All men are created equal," which women finally heard as "All human beings are created equal." Two hundred years later that idea is still sinking in and creating upheavals in the self-images of both women and men.

These enormous changes impact every family. We don't give ourselves credit for surviving them relatively sanely; we also don't offer ourselves or one another much compassion for our remaining, understandable disorientation. More often we bash each other for our different ways of reacting to this upsetting time.

Some of us accept or at least are resigned to death and birth control, the fragility of marriage, the industrialization of the family. We favor birth control, day care, easy divorce laws, court-enforced equality for women. We may not like abortion, but we see it as part of the historic tide. Like the pill in Massachusetts twenty years ago, we know it will be available no matter what the laws say, at least to those who are privileged.

Some of us are willing to accept only half the package—death control but not birth control, democracy but not equality, industrial growth but wives at home and men bringing home an occasional deer with a gun—not because we need the housewifery or the deer but because we need to feel like women and like men.

Most of us feel caught in the middle, wondering how others can sound so certain, wishing the loud, intransigent ones would just shut up.

I suppose all historic social changes look this messy. I guess it's human to turn our inner struggles into outer strife. But I do wish we could calm down and look at each other with some patience and sympathy. We are going through a hard couple of hundred years.

POVERTY AND
DEVELOPMENT

THE PROBLEMS of poverty and pollution are often set against each other. Produce what people need or protect the environment. Choose one or the other. To solve one problem you have to expect the other to get worse.

That choice just does not make sense to a systems analyst. It assumes the economy and the ecosystem are not only separate but competing entities, when of course they are integral to each other. Human needs can never be met from an impoverished environment; the environment can never be protected by impoverished human beings. Furthermore, no person with a conscience could live with either choice. Having lived among the poor, I cannot turn my back on them. Treasuring and depending upon the environment, I cannot trash it in the name of economic development.

So I refuse to be diminished by a false dilemma. I choose not either/ or but both/and. Provide the basic needs of human beings *and* sustain the health and fruitfulness of nature. Though I live in the richest land in the world, I can't ignore the problem of poverty; it is interlinked with population, the environment, and my own basic needs. Above all, it is interlinked with my own heart and soul.

A Decade Lost, When There Isn't a Decade
to Lose

IT HAS been possible to accept, temporarily, the grinding poverty of Third World nations because we think of those nations as developing, following in our footsteps, just a bit behind.

It has been possible to contemplate without despair the rapid expansion of Third World populations, in the belief that as they get richer, they will grow more slowly, as our own population and those of Europe have done.

Indeed, in most "developing" nations economic growth has been strong, and birth rates have been falling—until the 1980s.

During that decade, particularly in Latin America and Africa, nations became poorer. Birthrates rose. A new and frightening possibility presented itself. Maybe development isn't inevitable. Maybe "un-development" is not only possible but has actually begun.

UNICEF said in its 1988 *State of the World's Children* report: "In many nations, development is being thrown in reverse. . . . After decades of steady economic advance, large areas of the world are sliding backwards into poverty."

Maurice Williams, president of the Society for International Development, called the 1980s, "a lost decade for most of the developing world." Stephen Lewis, special adviser on Africa to the U.N. Secretary-General, refers to the "brutal and mindless 1980s."

In Latin America the average per capita income was 9 percent less in 1989 than it was in 1980. In Mexico wages went down by 30 to 40 percent. Some countries regressed economically to where they were twenty years ago.

In sub-Saharan Africa per capita GNP has declined by 16 percent since 1980. The World Bank projects that by 1995 Africa's output will

have climbed back up to 89 percent of what it was in 1970. Nigeria's
per capita income went from $800 to $380 between 1985 and 1987. The
minimum wage in Kenya declined 42 percent during the 1980s, and
the percentage of children stunted by malnutrition rose from 24 to 28.

UNICEF says that in the thirty-seven poorest nations of the world
spending on health services fell by 50 percent in the 1980s and on
education by 25 percent. In almost half the 103 countries of the Third
World the fraction of six- to eleven-year-olds enrolled in school has
decreased.

The foreign debt of the South climbed during the 1980s, from $400
billion to over $1 trillion. (In the same period world expenditures on
armaments went up by the same amount—from $400 billion to $1
trillion *per year.*)

In 1979 a net $40 billion flowed from North to South. Now, because
of debt repayments and capital flight, a net $85 billion each year flows
from South to North. From 1984 to 1989 Latin America paid out in
debt service $135 billion more than it received in capital inflows.

The World Bank and the International Monetary Fund are working
furiously on a plan to reduce Third World debt by $70 billion over the
next few years. That's 7 percent of the outstanding debt. It's not
anywhere near enough. Nothing being discussed by the Powers That
Be is enough. When roads, machines, and factories are not being
repaired, when children are growing up malnourished, ill, and unedu-
cated, when forests, pastures, fields, and mines are being plundered for
basic survival or to pay back debt, more than just a decade is being
lost. At least a whole generation is being lost—and maybe the opportunity ever to rise out of poverty.

Worst of all, because in most countries death rates are down but not
birth rates, populations are growing at nearly their maximum rate.
Between 1978 and 1988 the population of the Third World increased
from 3.2 to 4 billion, an increase three times the total population of the
United States. If these growing populations don't experience an improvement in their lives, their birthrates will not come down. They are
at the toughest point in the development process; the point where the
biggest push is necessary to get over the hump—and to keep from
rolling back downhill.

In case you think the development of the poor countries has nothing
to do with you, consider that Third World economic reversals have
cost the United States tens of billions of dollars a year in lost exports

and over a million jobs. We are pleading with Third World peoples not to level tropical forests, not to scuttle international treaties about the ozone layer and greenhouse effect, not to compete in our markets, not to sell us drugs, not to migrate across our borders, not to disrupt our airlines with their desperation and terrorism, not to form mineral cartels, not to disrupt the international monetary system. Our welfare is intertwined with their welfare in a hundred ways. One thing we can't afford is for them not to develop.

Trading Debt for Nature Instead of Nature for Debt

IT IS a tragedy, but not an accident, that the world's most threatened tropical forests are in some of the world's poorest and most in-debted countries. The very poverty of those countries destroys the forests. To pay at least the interest on their debts, governments encourage logging companies to take timber. Forests are turned into pasture for beef exports. Along the logging roads and into the overgrazed pastures spread settlers desperate for land, any land from which they might scratch a living. The forests don't stand a chance.

The World Wildlife Fund says that at the present rate of loss, the tropical forests may be gone in thirty years and with them at least 40 percent of earth's species of life.

The fact that tropical forests are priceless biological treasures does not thrill impoverished people or debt-burdened governments. Biologist Edward O. Wilson recently announced that he found forty-three species of ants (as many as there are in all of Great Britain) on a *single tree* in Peru. But exotic ants don't provide jobs or pay off the billions of dollars owed to foreign banks. So the forests go down, and the ants with them. An inexorable economic logic sacrifices nature for debt.

It was Thomas Lovejoy of the World Wildlife Fund who first saw a way to turn that logic around. In a 1984 *New York Times* editorial he proposed swapping debt for nature. Now several such swaps are under way.

To understand how a debt-for-nature swap works, you have to begin by understanding that Third World debt is bought and sold on the world market much like wheat or oil.

For example, a bank loans a country $1 million in exchange for a note promising repayment over a certain period at a certain rate of interest. The country starts missing interest payments or asks to stretch out the due date. There is an increased risk that the loan won't be repaid. Rather than run that risk, the bank may decide to sell the note for certain cash, though that cash will be less than $1 million, sometimes very much less. A million dollars of Costa Rican debt goes these days for about $170,000. If you want to hold $1 million worth of Sudanese debt, you can get it for about $5,000, which expresses, of course, roughly the probability of $1 million ever being paid.

Lovejoy's idea was that conservation organizations could obtain some of this devalued debt, by purchase or donation, and trade it for tropical forest protection.

For example, last year the World Wildlife Fund (WWF) bought a $1 million Ecuadorian IOU, due to be paid off in eight years. The price to the WWF was $354,000 (which it raised partly from a private philanthropist and partly from its own program money for Ecuador). The WWF arranged with the government of Ecuador to exchange that note for a bond worth $1 million in Ecuador's currency (sucres).

The sucre bond was given to Fundacion Natura, an Ecuadorian conservation organization, which will use the interest and eventually the principal for national parks—not to acquire parks, because many exist in Ecuador, but to train and pay park staff, guard perimeters, build research stations, set up nature education programs, and make the paper parks into real, protected parks.

Last year Conservation International bought $650,000 worth of Bolivian debt for $100,000. They exchanged it for an agreement by the Bolivian government to protect 4 million acres of forest and grassland adjoining the existing 334,000-acre Beni Biosphere Reserve in the Amazonian basin.

The World Wildlife Fund has just announced the largest debt-for-nature deal yet, worth $3 million. The equivalent in Costa Rica's currency (colones) will go to the Costa Rican National Parks Foundation. It will endow the Monteverde Cloud Forest Reserve and complete the purchase of land for Guanacaste National Park. Let's put that deal in perspective: The WWF mobilized in one stroke more money than it has been able to raise for Costa Rica over the past twenty-five years.

Third World governments willing to exchange debt for nature are not only ecological good guys; they are also getting a good deal. They exchange foreign currency obligations, which they must pay through export earnings, for domestic obligations to their own people, for some of their own priorities. Monteverde is not only a diverse and beautiful cloud forest; it is also the watershed above Costa Rica's largest hydroelectric plant. Restoration of Guanacaste is employing many local workers and bringing in a stream of international researchers eager to study one of the last remaining patches of American tropical dry forest.

Gatherings of conservationists these days are almost unrecognizable because the language is that of international banking and the sums are in the millions of dollars. But everyone knows that the nuts-and-bolts economic talk is actually about something beyond price. A Nature Conservancy brochure lays out what the real deal is. It says: Just $300 for each hectare (2.5 acres) of Guanacaste National Park "buys you all of (forever): 0.001 jaguar, 200 orchids, 0.5 parrot, 20 toads, 25 spiny pocket mice, 0.04 anteater, 100 vines, 0.03 spider monkey, 400 dung beetles, 0.01 muscovy duck, 0.00029 volcano. All purchases will be held for your on-site inspection by the Costa Rican National Park Service."

Hunger and Surplus Rise and Fall Together

THE SIMULTANEOUS occurrence of hunger and of mountains of surplus food is obscene. It's also expensive. And it's persistent, in spite of hunger-reduction and surplus-reduction policies of governments the world over.

In a typical year the United States, Japan, and Europe spend $100 billion to protect their farmers against low prices caused by agricultural overproduction. Half that amount goes to farmers; the rest goes to bureaucracy and to storage of grain, butter, and milk.

The surplus grain stock of the European Community in 1984 was enough to feed fifty times the combined populations of Ethiopia and the Sudan that year. India had 25 million tons of unsold grain, enough to feed all the hungry people in India. But it was not feeding them: it was rotting in storage.

The problems of hunger and glut persist, says Ferenc Rabar, because we attack them as separate problems. They are not separate. They both arise from the normal operation of the world food system. They can't be solved separately, he says, but they can be solved together.

Rabar is a Hungarian economist who founded in 1976 a research group at the International Institute of Applied Systems Analysis (IIASA) to learn how the world food system works.

With remarkable international cooperation, teams in the United States, East and West Europe, and many Third World countries constructed computer models of their nations' agricultural production, consumption, and government policies. These models were then linked at IIASA into a simulated world trade network.

The resulting megamodel reproduces the pulls and tugs of the actors in the world market, each constantly adjusting to the others. If the Japanese eat more meat, the price of American corn goes up. If the Russians have a good harvest, the price of wheat goes down and Nigeria can afford more imports. If the United States floods the market with cheap corn, Europe puts up trade barriers. And so forth.

The model shows that the rational actions of farmers, consumers, and governments produce the conjoined problems of hunger and glut. Those problems are produced relentlessly, under many different changes tested in the model system. For example:

- A drought in India increases hunger in India. A drought in Kansas increases hunger in India even more.
- If a Green Revolution raises agricultural production in the Third World, much of the increase is exported to the West, gluts go up everywhere, and hunger is reduced only slightly.

- If Americans eat less meat, world meat and grain prices fall. Moderate-income consumers eat more meat for a while, until falling prices put farmers out of business and meat and grain production go down. There is only a small decrease in the number of hungry people in the world.
- If 50 million tons of free grain are dropped into the world market each year—like manna from heaven—the manna does not get to the hungry. Most of it is bought by rich nations to feed animals grown for meat. And the free grain lowers grain prices, which then wipes out farmers.
- If all the agricultural trade barriers of the developed countries are removed, food prices rise, farmers are better off, the rich countries' economies grow faster. But because of the higher prices poor countries import less food. Gluts decrease somewhat, but hunger is hardly affected.

Hunger is impervious to all these changes because they are changes in markets, and markets are irrelevant to hunger. People who have no money are bypassed by markets; they are simply not hooked into the world's food distribution system. In rich nations gluts pile up because people there are already buying all the food they want. Poor nations, where food is needed, don't have the money to buy it. More production, lower trade barriers, manna from heaven, all these changes cause market readjustments, but they don't bring the hungry into the market.

A *really* free market, free for labor too, could reduce both hunger and glut. If migration barriers are removed in the IIASA model, people go freely to land and jobs and grow or earn enough to feed themselves. Surpluses go down. Farmers and nations prosper. According to Rabar, that would never happen in the real world. Migration barriers protect not only the high wages of domestic laborers but also the cultural identities of nations.

But there is another successful policy. It combines three changes that make little difference separately but combined eliminate both hunger and glut:

- Worldwide agricultural trade liberalization
- A foreign aid program that transfers income from the rich countries (0.5% of their GNPs) to the countries where hunger is greatest

- Investment of that aid in development projects that generate income for the poorest people

Under this combination of policies, the entire world economy grows faster. Amazingly, *the economic gain in the rich countries is more than enough to finance the aid flows.*

When Rabar showed me that calculation, I found it hard to believe. I have not been conditioned to expect win/win solutions. But this one was worked out painstakingly, country by country, price by price, year by year, and agreed upon by economists from twenty nations.

When you think about it a while, you can see it makes sense. Instead of handing billions of dollars to farmers not to grow food, why not use those dollars for real economic development, so the people who need food can enter the market and buy it for themselves?

The Desert Blooms in Auroville

AUROVILLE, NEAR Pondicherry in the province of Tamil Nadu in South India, is one of the world's strangest communities. It consists of 2,000 acres of farms, forests, and settled areas, scattered over a desiccated plateau, interspersed with traditional Tamil villages. About one-fourth of the 600 inhabitants of Auroville are Indian, the rest come from twenty-five nations, most of them European. Auroville is not only a model of an international community; it is a model of how to create sufficiency and development in one of the most desolate landscapes of the world.

The idea of Auroville began decades ago when a few Europeans came to India to seek the wisdom of the East. They were disciples of Sri Aurobindo, one of India's many spiritual leaders. He died in 1950 and left his followers with a vision of a community founded on the principle of unity—unity of human beings with each other and with the earth. Sri Aurobindo left no blueprint for this community; the followers had to work it out for themselves.

They started in 1968 in one of the most discouraging parts of India. The coast of Tamil Nadu has long been deforested, severely over-grazed and overcropped. Every year there are six months of drought, during which the wind blows away the topsoil and the sun bakes the earth into a pitted moonscape. Then there is the annual monsoon, during which heavy rains rush off the land, washing away any soil that remains. The only plants that survive are thorny scrub and the tough palmyra palm, which makes weird patterns against the blazing sky, each tree bearing only one leaf—the villagers lop off all the rest to feed the goats.

Here the Aurovillians proposed to farm. They had to start by accumulating soil and water. They adopted from the farmers around them the idea of "bunding"—constructing earth barriers to catch the water running off the fields, so it has time to sink into the groundwater. All Indian farmers know about bunding, but few have the strength, foresight, and community cooperation the people of Auroville put to the task. With hand labor and hand tools and over many years, Au-rovillians created bunds, and more bunds, and still more bunds.

As the water was redirected into the earth, depleted wells began rising in Auroville and in all the villages around it.

Around the bunds they planted trees to slow the wind and hold the soil. The first trees had to be hardy ones that could stand the desert environment. Many of them were nitrogen-fixing, to build up soil nutrients. Thorny hedges protected seedling trees from grazing animals. Later, as shade from the first trees cooled the land and as dropped leaves built the soil, food-producing trees were able to grow—mango, cashew, jackfruit, lime, papaya, banana.

Over seventeen years Aurovillians planted more than a million trees. With the return of the trees, other kinds of life reappeared, birds and butterflies that hadn't been seen in the area for years.

Soil was built by persistent composting, by green-manuring, by returning organic wastes to the land. One by one small patches of garden and field were created. When Auroville started, it could supply none of its own food. Now, with a much larger population, it supplies 70 percent of its own needs and sells surplus milk, fruits, and vegetables. It uses no purchased fertilizers or pesticides.

Auroville, like its neighboring villages, has never had much cash or capital. Its success has depended on one resource that its neighbors share—labor—combined with another resource that is transferable—Western knowledge, organization, and, above all, self-confidence. The Tamil neighbors have been watching Auroville. Some of them have been working and living there. Now they are beginning to make their own bunds, to dig holding ponds, to take home tree seedlings and new vegetable varieties, to make compost. Many of them have also learned marketable skills in Auroville's craft shops.

The government of India now contracts with Auroville to manage nearby government lands, to undertake tree-planting projects, and to run training sessions in village crafts and marketing cooperatives. Auroville, having just established its own economic viability, is beginning to think about spreading more widely the lessons it has learned.

All this has taken almost no outside funding and seventeen years. (One Aurovillian reminded me, "Seventeen years is nothing in the time-scale of India.") It has directly benefited only the residents of Auroville and their immediate neighbors. It has turned a few thousand acres of desert into lush forests and gardens.

Is it a model for anywhere else?

Probably not, in exact detail, but I think it suggests some useful general lessons. The first is not to give up on any land, no matter how harsh and forbidding. Care and hard work, understanding and wise use of natural forces, can create miracles.

There is a second lesson in the humility of the Aurovillians. They did not go to South India to "develop" anyone but themselves. They were open to the knowledge of the local people, and they also used the strengths of their own Western cultures. Inadvertently, and with many struggles (I have not said enough here about their struggles) they learned to solve the on-the-ground problems of the area, with the tools of the area.

Yet another lesson is patience. The restoration of land and of confidence requires decades of commitment. The Aurovillians are not doing a hit-and-run goodwill project. They are building their own community, to which their own futures are tied. They are in it for the long haul.

Development took hundreds of years in the West, and it did not start with a degraded ecosystem or a demoralized people. Development will take a long time in Asia and Africa. We need to be steadfast partners, however long it takes.

Sarvodaya Shramadana: The Awakening of All by the Sharing of Our Gifts

IN 1958 a high school teacher in Sri Lanka took some of his students on a "holiday work camp." They spent eleven days in a low-caste village, digging wells and latrines, clearing the land, and learning about the life and the problems of the people.

That was the beginning of the Sarvodaya Shramadana movement, which eventually reached 6,000 of the poorest villages of Sri Lanka—one-fourth of all the villages in the country. Sarvodaya means "the awakening of all." Shramadana means "by sharing our gifts."

The Sarvodaya villages are definitely awakening. Mothers are starting cooperative sewing projects; prisoners are clearing playgrounds for children; nurseries and kindergartens are beginning; irrigation canals are being dug; weaving and soap-making industries are starting. All this is happening with almost no outside support, just from the power of the people themselves and from the vision of that high school teacher, whose name is A. T. ("call me Ari") Ariyaratne.

A village joins the movement by inviting a team of Sarvodaya organizers to visit (in the mid-1980s there were 27,000 full-time organizers). The team calls a meeting in the local temple, church, or mosque and asks what the village most needs. Then the work begins.

"You say you have waited two years for the government to clean that canal? You can keep on waiting while your fields bake. But where is your own power? Your power is not in Colombo; it is in you, in your head and hands."

"How can we clean the canal? We have tools, but no earth pans." "Is there a substitute for pans?" "Yes, sheaves of leaves." "How many people to do the job?" "Two hundred, working four days." "How many volunteers—can each one bring one? Right, who will feed them?"

One rich landowner offers food for two days. "Who will feed one other by sharing one day's meals?" The hands go up. The canal is finished, not in four days, but in the afternoon of the first day.

The project shows the people what they can do together and how to organize themselves. "We build the road, and the road builds us." With that motto Sarvodaya projects have built in one year eight times as much roadway as the government, at one-eighteenth the cost.

Schools and clinics have been built, and better houses. At the same time people are encouraged to grow in spiritual and communal directions. There are times for music making, for celebration, for meditation or prayer, for conflict resolution. "We wouldn't want to make better houses with unawakened people in them."

Sarvodaya has set forth its agenda in its list of the ten basic needs of all people. It's an interesting list; not exactly what we rich Westerners might imagine poor people would set forth as their most urgent needs. It reads, in order:

- a clean, safe, beautiful environment
- a clean and adequate supply of water
- enough clothing
- a balanced diet
- basic health care
- communication
- education
- fuel
- simple housing
- spiritual and cultural needs.

Ariyaratne says, "The West has given us the 1,2,3,4 model of development—1 wife, 2 children, 3 bedrooms, and 4 wheels. We don't think that model works for the poor or the rich. We believe that development is an awakening process within individuals, within families, within communities, within the entire humanity—spiritual, moral, cultural, political, and economic awakening. Development should not be reduced to filling your house with goods and gadgets and becoming a prisoner of what you possess."

Ari is both a practical organizer and a spiritual leader in the tradition of Gandhi but with a Buddhist vocabulary. His voice is gentle and his manner loving, and, like Gandhi, he has an impish sense of humor. The government alternately fears him and works with him. He refuses to join the government directly or to take funding from it, but he will train government ministers to become workers in the movement.

Nonviolence and religious tolerance are the essence of the Sarvodaya movement; it works in Buddhist, Hindu, Christian, and Muslim communities. And yet, just as Gandhi had to witness terrible outbreaks of Hindu-Muslim violence among the Indian people, Ari has had to live through bloody Buddhist-Hindu strife in Sri Lanka. During the worst of the riots, Sarvodaya workers housed and fed thousands of Hindu refugees and tried to calm the people. They held retreats to meditate and to get to the source of racial hatred. At one point Ari walked across the entire country, appealing as best he could to the nobility in each person, rather than to the evil.

He takes the civil violence hard. "When I cry for the suffering of the poor," he told me once, "I cry in public. When I cry from disappointment with the poor, I cry alone."

And, like Gandhi, he goes on talking of peace and empowerment. "We talk of first, second, and third worlds, but we find all those worlds in all countries. Our problems and yours are common problems, and we have to find a common philosophy. I dream of committed people everywhere sharing a universal awakening, seeing that the well-being of humanity is interlinked. Wherever people suffer, let us share, and let us have not one leader, not ten or a hundred leaders, but thousands of leaders, everywhere."

What Would the World Be Like If There Were No Hunger?

ABOUT ONE out of every eight persons on earth is chronically hungry. The death toll from hunger is equivalent to 100 fully loaded jumbo jets crashing *every day*. Most of the victims are children.

Suppose we put a stop to that devastation. What would happen? I don't mean what process would we go through to end hunger. I mean what would it be like when the job was done. How would it feel to live in a world where no one had to live or die in hunger?

Once in a seminar I asked that question of people who devote their lives to ending hunger. They work in aid programs, in agricultural research and extension, in relief agencies. You would think they must have some image of the goal they are working for.

But if they do, they aren't eager to talk about it. My question set off all sorts of mental short circuits.

One person said hunger will never end, so it's nonsensical to talk about it. Another said that he only needs to think about how we get there, not what the world will be like when we do. Some people became angry and resisted my question. To glimpse even for a moment a world without hunger just deepened the hurt of living in a world where hunger is a daily reality.

Some of the hunger workers told me they distrust visions of ideal worlds—throughout history visions have caused nothing but trouble, they said. And others admitted that they have a guiding vision of the end of hunger, but that it's personal, something that makes them feel childlike and vulnerable when they bring it out in the open.

And then there were those who hesitantly started talking about their visions of the end of hunger.

One of them said, "In a world without hunger, if you took a poll and asked kids what they want to be, being a farmer would be at the top of the list, not the bottom."

Another said, "There would be education, health care, and clean water for everyone. Population would stabilize because people would understand about health, about their bodies, and about what it means to give and nurture life."

After the first few people broke the ice, the visions began to pour out, as if they had been waiting a long time for release. For instance, "When I picture a world without hunger, I see lightness, by which I mean a human lightness, playfulness. Human beings in that world will find it easy and joyful to care for each other. No one will have to worry about anyone on the other side of the world because people will be taking care of each other near at hand."

Another said, "The air would feel new and fresh to me, the meals would taste incredibly good, my home would feel comforting in a storm, because for the first time in my life, I would know that everyone was clothed and fed and sheltered. I would be able to experience my life without being split apart by guilt and sadness."

And another burst out, "When the last person is fed, there will be a qualitative shift. People won't rush by each other. When hunger is ended, instead of single notes beaming out of each of us, there will be a symphony—we will see that each one of us is *fascinating*!"

Whenever a conversation gets that poetic, you can count on some realist to bring the mood back down to earth. "You know," one person said, "we've made significant inroads on hunger in the United States, but I don't sense any lightness or symphony. Why not? Is it because of the half billion over there who are still hungry? Or are you visionaries all wet? Maybe when hunger is eliminated everywhere, it will just be like the United States now, with all the stresses and strains."

Maybe so. Maybe it's dumb to go around talking about a world without hunger.

But I keep doing it. I think it's essential. Behind every decision to fund a foreign aid bill or start a community garden, to give to an aid organization or breed a high-yielding strain of wheat, to take a stand for justice or not to, there is some kind of vision. Visions alone don't produce results, but we'll never produce results without them. The absence of vision is one of the main reasons there is still hunger on this earth.

If we think that the end of hunger will be a great, grim sacrifice, that it will take away our own comforts or privileges, that it will require worldwide regimentation, that the only beneficiaries will be those who are hungry, or that the world wouldn't be very different, what would be the point of ever working to end hunger? If, on the other hand, we saw that ending hunger would be easy, that it would be joyful, that it would allow hundreds of millions of people to become productive partners, that all our lives would be enhanced immeasurably, then imagine what might we do.

Imagine, really imagine, a world with no one living in dire need. Think how it would feel to be part of a society that had taken on and solved, permanently, the problem of hunger. I think the person who has come closest to expressing how it would be is the Spanish poet

Frederico Garcia Lorca (1898–1936). His vision is the one I carry around with me:

"The day that hunger is eradicated from the earth there will be the greatest spiritual explosion the world has ever known. Humanity cannot imagine the joy that will burst into the world on the day of that great revolution."

HEALTHY FARMS, HEALTHY FOOD, HEALTHY LAND

THOUGH I didn't grow up on a farm, I've been attracted to them all my life. When in 1972 I finally came to buy my own home, it was a farm. My psychological roots grew instantly into its cold, rocky soil. I have tried several times to leave it, reasoning that I could write more if I didn't spend so much time shoveling manure, that I need to be where the political action is, that I'm not a very good farmer anyway, that New Hampshire is a terrible place to farm. But I've always come back. Something deep in me needs to be attached to a farm.

So nothing makes me sadder than the state of the world's farms, economically and environmentally. The economic transformation of family farms into industrial factories, the technical transformation of biological processes into chemical frenzies, the loss of soil, the poisoning of the countryside, the bankruptcy of farmers, none of these is right. We don't need to endure all that to feed the world, and if we keep it up, we won't be able to feed the world much longer.

I am an organic farmer. I hang around other organic farmers. I have paced over the soil, smelled it, checked the crops on dozens of organic farms, large and small, on three continents. I know organic farming works. Sometimes I wonder why it's taking so long for other people to figure that out.

Sentimentality and the Family Farm

T HE LARGE farms of this country are doing fairly well. Most of the small farms are supplemented by outside income. The current epidemic of farm foreclosures is affecting primarily middle-size farms, owned and operated by single families dependent almost completely on farming for their income.

So what? The farm *land* is not going out of production. Most farms that go up for auction are bought by expanding neighbors. Even if the last family farm disappears, America will go on eating and exporting. Maybe it's time to get over our nostalgia and welcome the industrialized megafarm of the future.

I asked Congressman James Jeffords of Vermont how he defends the family farm to his colleagues on the House Agriculture Committee. "There's nothing I can say. The only thing that works is pictures." Pictures? "Pictures of farm families, standing on their farms. When they see that there are 80,000 *families* going out of business this year, honest people who have worked hard, then they know that something important is going wrong."

I was disappointed in that tactic. I wanted a hard-nosed reason to support the family farm. Sentiment is no proper basis for public policy. But the more I thought about it, the more I realized that farm policy is already based on sentiment, always has been, always will be, and that's not bad, as long as the sentiment is right.

Wendell Berry, author and farmer, identifies the two opposing sentiments in the farm debate as "exploitation" and "nurture."

"The standard of the exploiter is efficiency," he says. "The standard of the nurturer is care. The exploiter's goal is money, profit; the nurturer's goal is health. . . . The exploiter wishes to earn as much as possible by as little work as possible; the nurturer expects, certainly, to

95

Figure 3

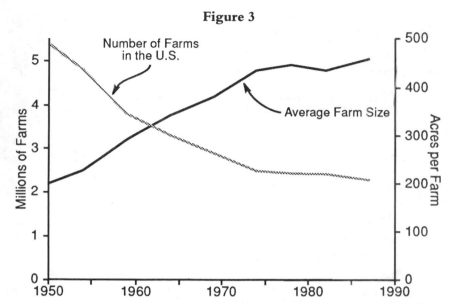

Size of farms has risen; number of farms has fallen. The apparent leveling-off of farm size since the 1970s has come as very large farms and very small (part-time) ones have proliferated, while mid-sized family farms are failing.

SOURCE: USDA

have a decent living from his work, but his characteristic wish is to work as well as possible. . . . The exploiter thinks in terms of numbers, quantities; the nurturer in terms of character, quality."

We who speak for the family farm think that farms should be small enough to *care* for. Though there are certainly small farmers who don't care enough and large farmers who do, we argue that at some size even the best of intentions and technologies cannot compensate for human attention spread too thinly over too many acres.

We worry about the quality and quantity of jobs. Farm jobs seem to be the only ones the nation does not care about losing—perhaps because we think of them as the back-breaking, demeaning work of the previous century. They are in fact entrepreneurships, positions of responsibility. The people who hold those jobs do not leave them willingly. They do not move on to something better.

We farm sentimentalists care about the health of families and communities. My farm neighbors work together for a common enterprise. They are "families" in the most solid sense of that word. The kids are too busy raising 4-H calves to be bored and dangerous. The families are rooted in the town, active in its governance, present at its celebrations. The difference between a family-farm town and a company town is real and important.

Finally there's Thomas Jefferson's issue of democracy and power. Something in us all fears the agglomeration of productive power. We do not like to feel at the mercy of distant corporations for our daily needs. There's a social strength when the means of production are widely owned; there's a weakness when they're centralized.

The appeal of the opposite sentiment, that of the exploiter, is, as far as I can see, in the challenge of accumulating money and power, land and control. The discipline and thrill of competition. Wondrous technologies Release from the tyranny of labor and of nature. Freedom for nearly everyone to do something else, while a few agribusinesses provide our food, cheaply and comfortably.

The vision on that side goes something like this (courtesy of the U.S. Department of Agriculture):

"Attached to a modernistic farmhouse, a bubble-topped control tower hums with a computer, weather reports and a farm-price ticker tape. A remote-controlled tiller-combine glides across a 10-mile-long wheat field. The same machine that cuts the grain prepares the land for the next crop in a single pass. A similar device waters neighboring strips of soybeans as a jet-powered helicopter sprays insecticides."

It goes on, but you get the picture.

On the other side, the vision is far less flashy. Here it is expressed by a Kentucky farm woman:

"Well, we stayed with farming because we like farming. We like the dirt. We like to see things grow. Then, we're our own boss. We can quit in the middle of the day if we want to. We can work until midnight if we want to, and it's a free life and a good life. I think contentment of heart goes a long way in lifting up the social life of our world and being happy with what you have and not reaching out and grasping for being a millionaire, and counting your blessings, living with your family, and appreciating good things."

In each of us there is some exploiter and some nurturer. Both attitudes permeate American history and culture. The tension between the two is the tension between Alexander Hamilton and Thomas

Subsidies would help, but only if they were limited to small and middle-sized farms and if they eased a farmer's debt and supplemented his income, instead of increasing the size of his operation.

A Farm Policy That Might Work

HOW COULD a farm bill be written to guarantee the survival of the family farm?

Well, to begin, it ought to have that as its goal. The main problem with our current farm policy is that it has too many goals. We are trying to make farm prices high for the farmer but low for the consumer, to produce all we can for export but reduce surpluses, to render humanitarian food aid but use food as a foreign policy weapon, to conserve farmland but make it produce more and more, to encourage young farmers but weed out inefficient ones, to stabilize prices but allow the market system to work, to appease special interests and get the administration reelected, and, oh yes, to feed people.

Unless we sort out these goals, our farm policy will continue to be chaotic.

It isn't that hard to recognize the self-serving goals on that list and separate them from the true societal ones. I'd suggest, for a farm policy goal, that we try to ensure that our nation has numerous independent farmers, each operating at a scale that encourages good stewardship, good communities, and efficient production, and each earning a decent living. We wouldn't need to worry after that. They'd produce enough food.

How to achieve that goal?

First, we need some measures to help the family farm weather the current storm. An emergency loan program, limited to medium-sized and small farms, would help them hold on until we could do something more permanent. If we could bail out Chrysler, we can bail out the people who feed us.

Second, we should sort through our current farm policies and remove all the biases that favor large farms As a partial list we should do the following:

- Eliminate the numerous ways farms can be used as tax shelters.
- Exclude farmers with high off-farm income from eligibility for commodity payments.
- Rewrite inheritance taxes to allow moderate-size farms to stay in a family.
- Limit commodity payments to yield no more than a fair income for each producing family.
- Restrict federally subsidized loans only to family-sized farms.
- Allocate federal research and extension resources to the clear benefit of small and medium farms, and make a special effort to extend new technologies to those farms.

Corrections like these might be enough to stabilize the family farm, but I doubt it. They remove some unfair advantages of large farms, but they leave in place the competitive mechanism that forces farmers always to expand. We need to find a way of preserving the advantages of the market—the push toward efficiency, the responsiveness to consumer demand—but to relieve the market's constant pressure to produce more. And we need to protect the farmer from the turbulence of the speculative land market.

The place in the world that has come closest to doing all that is Scandinavia. In Denmark and Norway prime agricultural lands are strictly zoned for agriculture only—no exceptions, no arguments. The land is still privately owned, still bought and sold on the market, but since it can be bought only by farmers, its price is determined only by the farmers' own economics; it reflects the return expected from agriculture, not from shopping malls. Young farmers don't have to bid against speculators for land. If we had that policy here, we would have eliminated all the technical farm bankruptcies that occurred in the 1980s when an overblown speculative bubble finally burst and land prices fell.

In Norway one cannot own farmland unless one is willing to live on it and farm it personally. No absentee landlords, no corporations, no tenants, no tax shelters. Furthermore, no one is allowed to farm without an agricultural degree. (The universities are free to all citizens.) That guarantees farmers who know about new technologies and about soil conservation.

Though I think these measures would be beneficial, still they do not directly interrupt the market treadmill. There is one astonishingly simple (and politically unthinkable, at the moment) way to do that. Limit the size of farms. Just put on a cap. Define some upper limit beyond which a farm cannot grow. Set it high enough to capture economies of scale and a decent farm income, low enough to ensure healthy land and communities.

While the squawks die down (Communist! Socialist! Interference with our freedom! Interference with the market system!), let me elaborate. The limit should vary by crop and land type. It would be most effective to set it not by number of acres or dollars of income, but by the amount of each commodity that each farm would be permitted to sell—so many hundredweight of milk or bushels of grain.

A limit in real commodity units would give the government a way of dealing with overproduction. It would also give farmers maximum freedom to experiment with different management schemes to produce the limit at lower cost. Since the total amount produced would be fixed, any decrease in costs would mean an increase in farm income, not farm size.

Limits would have to be changed occasionally, at first to find their proper level, later to reflect changes in technology or in the world market. Setting and changing them would involve real political infighting—just like what we have now. But if we could keep the process honest (a big *if*), I suspect that a farm limit would eventually eliminate all need for subsidies because farms would be profitable and stable without outside help.

Please note that this scheme would still keep the farm economy very much capitalist. Farms and produce would be privately owned and bought or sold at will. The market would be constrained in overall size, however. It would no longer be the master of the system, but its servant.

What gives me the greatest confidence in this proposal is that I have heard it suggested many times by farmers themselves.

Where Do All the Pesticides Go?

E VERY YEAR American farmers apply 1.3 million tons of pesticides to their fields.

When pesticides are sprayed by airplane—and 65 percent of them are—less than half the chemical hits the target field. The rest disintegrates in the air or falls somewhere else. Of the pesticide that does reach the field, far less than 1 percent makes contact with a harmful insect or weed.

American farmers pay $4.1 billion a year for these chemicals, 99 percent of which do not hit home. For most farmers, looking only at their own balance sheets, that investment is sound in the short term.

But if we could trace where those millions of tons of chemicals go and count up all the costs of their presence, we would not be impressed with the economics of pesticides.

Some pesticides end up in groundwater. Citizens of the Central Valley of California, surrounded by some of the most heavily sprayed fields in the country, are finding their water seriously contaminated. Wells on Long Island, polluted with the insecticide Temik, will probably not become safe for drinking for 100 years. The costs can be counted in the importation of bottled water, in the latent cancers in children, and in the shattered peace of mind of townships.

Some pesticides kill species that are not pests, creating various kinds of ecological havoc. A Nebraska farmer told me that the songbirds disappeared from his farm, as well as the bull snake, which kept the mouse population in check. He stopped using pesticides and gradually the wildlife came back.

Sometimes pesticides actually create pests by wiping out species that naturally keep pest populations under control. Cotton farmers used to apply DDT and dieldrin to control the boll weevil. The pesticides killed off the natural predators of two other insects—the tobacco budworm and the bollworm—which until then had never been a problem. Without their predators they multiplied and became major cotton pests.

The creature most in danger of being accidentally zapped by pesticides is the farmer. A recent study of Kansas farm families has revealed cancer rates far above the national average. The cancers are correlated with use of the herbicide 2,4-D.

California registers 200–300 serious pesticide poisonings of farmworkers every year, and the California Health Department estimates that only 1 or 2 percent of poisonings are ever reported. These incidents are acute toxic exposures only; the slow accumulation of cancers and birth defects is hard to follow, especially since in California many of the workers are migrants.

Epidemiologists guess that over 300,000 farmworkers in the United States suffer pesticide-induced illnesses every year. There is no federal protection guaranteeing farmworkers a safe working place, as there is for industrial workers under the Occupational Safety and Health Act (OSHA).

Even the pesticides that strike home and eliminate the intended pest can cause problems. A few genetically resistant individuals may survive. They are the ones that go on to produce the next generation, which inherits the resistance. The pesticide industry is in a race to see whether new pesticides can be invented faster than the pests become resistant to the old ones.

After the boll weevil became DDT-resistant, farmers switched to methyl parathion, continuing the DDT to control budworm. By 1965 the budworm was resistant to DDT, to other chlorinated hydrocarbons, and to the new carbamate pesticides. A few years later it was also resistant to methyl parathion. Farmers were spraying chemical combinations in increasing quantities ten to twenty times per season but still were losing their crops.

More than 400 insects and mites have become resistant to pesticides in this country. Over 150 microbial pathogens are resistant to fungicides, and at least fifty weeds are showing herbicide resistance. The more the chemical tools are used, the sooner they generate the resistance that ends their own effectiveness.

The manufacturing plants that make pesticides can be deadly neighbors, as was demonstrated in Bhopal. Even when they don't malfunction, they are a major source of toxic waste.

In short, the $4.1 billion tab to the farmers is only the beginning of the real cost of pesticides.

But then there are the benefits: the billions of dollars of income and the hundreds of thousands of jobs in the agrichemicals industry, the ability to produce food in abundance, and, for the farmer, an arsenal of tools against the invading hordes that threaten to munch away his livelihood.

Weighing the costs and the benefits is not easy, which is why we still have pesticides and still go on arguing about them. But looking at the whole picture, one has to wonder whether there isn't a more intelligent, less expensive, and less clumsy way of coping with pests.

Pesticides, Residues, Diets, and Kids

IN 1988 President Reagan signed into law a set of revisions to the Federal Insecticide, Fungicide, and Rodenticide Act (known as FIFRA to its friends and enemies). The revised law has been nicknamed FIFRA-Lite by environmentalists, not because it weakens pesticide regulations—in fact it strengthens them—but because it still falls so far short of what is needed to protect the groundwater, the environment, the population, and especially the kids.

Under both old and new FIFRA the Environmental Protection Agency (EPA) is charged with being sure that the public comes to no harm from the use of 600 active pesticide ingredients and 1,200 "inert" ones sprayed on 376 agricultural commodities by 2 million farmers. That charge has always been and still is impossible.

The new FIFRA authorizes $110 million in federal funds over the next nine years and assesses chemical companies $150 million to review the hundreds of pesticides now in use that have never been properly tested for their health effects. To put that corporate $150 million in perspective, it is about half a cent out of every dollar of pesticide sales the companies will make, assuming prices and quantities stay constant.

FIFRA-Lite also removes the crippling requirement that the EPA compensate companies for inventories of banned pesticides (for some chemicals the amount of compensation would exceed the EPA's entire annual pesticide budget).

These are important and hard-won gains. But the new FIFRA does not protect groundwater supplies or farmworkers. It does not prevent U.S. companies from marketing banned pesticides abroad. It also does not address the rising concern about pesticide exposure of children.

I first got worried about the children four years ago when Professor John Wargo moved to an office across the hall from mine at Dartmouth College. He and two graduate students, Ross Brennan and Stewart Dary, were working on a project for the National Academy of Sciences, converting EPA data on pesticides to a microcomputer.

The buzz of conversation across the hall about Alachlor and Benomyl and tolerances interested me, and I kept asking questions. Slowly I began to understand what the EPA goes through to decide how much Atrazine we can eat with our corn oil, how much Alar with our apples.

According to FIFRA new pesticides (and, eventually, all old pesticides) must go through a battery of tests to determine their short-term toxicity and their long-term carcinogenicity. Toxicity is measured by feeding the chemical to laboratory animals in increasing doses until it has killed half of them. Tests of internal and external exposure over periods up to two years determine whether the pesticide causes tumors or birth defects.

The test data are submitted to the EPA. They are the base of scientific observation upon which the whole regulatory process is built. Already, however, some large assumptions have been made. *Assumption 1. The results have been interpreted correctly.* There was a big fight, for instance, about whether lesions caused by Dicofol were cancerous—it was decided that they weren't. *Assumption 2. The test lab is honest.* One of the big ones, Industrial Bio Test Labs, was discovered not to be.

The EPA must convert the animal tests to permissible human exposures. To do so, it must string together a long chain of even more uncertain assumptions.

First the EPA determines a No-Observable-Effects Level (N.O.E.L.)—the daily dose that produces no visible ill effect on test animals. *Assumption 3. No observable effect means no actual effect. Assumption 4. There is some safe level of exposure.* No assumption has been more fought over than that last one. It may be that carcinogens cause tumors even at very low concentration. Federal law has been inconsistent, permitting no carcinogens in processed foods but "safe" levels in raw foods. In fact, the EPA permits carcinogenic pesticide residues if they pose a "negligible risk" of inducing cancer.

Next, the EPA has to convert the N.O.E.L. from mice to men. *Assumption 5. Humans are 10 to 10,000 times more sensitive to pesticides than are test animals.* The EPA takes the N.O.E.L., divides it by a number between 10 and 10,000 (exactly what number is a matter for regulatory judgment), and calls that the Acceptable Daily Intake (ADI). For carcinogens it sets the ADI at what it assumes will cause no more than one cancer in 1 million people over a lifetime of exposure.

I can find no one who can describe exactly how the ADI from toxicity tests and the ADI from carcinogenicity tests are combined to determine an overall acceptable public exposure. There is no visible rationality in the final amounts of pesticide residues permitted by the EPA; there is no correlation between permitted amounts and either the toxicities or the carcinogenicities of the chemicals.

The apparent irrationality may be due to regulatory lag. The test data don't come in all at the same time (carcinogenicity tests take much longer). Many older chemicals have not been thoroughly tested. New findings keep turning up. The ADIs are constantly being revised, even as the chemicals go on being used in the fields and the foods of the nation.

Out of the chaos the EPA somehow determines, let's say, an ADI of 0.05 milligrams of Captan per kilogram of body weight per day. Captan is a fungicide used on apples, tomatoes, peas, sweet corn, onions, beans, squash, carrots, and oranges. To be sure that you don't get more than that acceptable 0.05 milligrams in your total diet, the EPA has to know how much of each of those foods you eat.

Assumption 6. We all eat the average American diet. The EPA has survey data on what a sampling of Americans eat. It averages us all together, you and me and your finicky two-year-old and my vegetarian grandfather. This, as we shall see in a moment, is the assumption that most threatens the children.

Given the average diet, the EPA sets *tolerances* for each chemical and crop—so much Captan residue can be tolerated on beans, so much on corn. Altogether the EPA has established 8,444 tolerances. For each one it sets forth *field procedures* specifying the maximum amount of each chemical a farmer can use on each crop and what period must elapse between application and harvest. The Department of Agriculture has the responsibility to publicize the field procedures and monitor compliance on the farms—another impossible task.

More assumptions have crept in here. *Assumption 7. The public's only exposure to a pesticide comes through food ingestion. Assumption 8. When several pesticides are present in the diet together, they do not do more damage than they would separately. Assumption 9. Farmers can and will obey field guidelines. Assumption 10. Obeying field guidelines results in tolerances being met.*

Fewer than 1 percent of commodity shipments are sampled by the Food and Drug Administration to test for compliance. They are not tested for all pesticides or even for some of the most common or most carcinogenic ones. From 5 to 8 percent of imported foods are found over tolerance, and 2 to 4 percent of domestic ones. Many crops tested have residues far below tolerance—for the chemicals that are tested. Some farmers (about 1% of all the farmers in the nation) use no pesticide at all.

My colleagues across the hall put the data for toxicities, N.O.E.L.s, ADIs, average dietary exposures, and tolerances on a computer spreadsheet. Then they asked "what if" questions. What would be the effect on cancer risk if Metolachlor is substituted for Alachlor? What if we define "negligible risk" as one cancer in 10 million instead of 1 million? What if some subsets of the population don't eat the average American diet?

That last question led to the children. Growing kids eat twice as much food per unit of body weight as adults. They eat simpler foods with less variety, much more milk, basic grains, and fruits. Average American apple juice consumption is 0.22 grams per kilogram of body weight per day, but the average nonnursing infant eats 3.46 grams per kilogram per day—more than ten times as much.

Wargo, Brennan, and Dary started by checking tolerances for eighteen commonly used fungicides against the average diets of children. They found that children one to six years old could be exposed to more than the acceptable daily intake of 10 out of those 18 fungicides. Some of the calculated overexposure levels were enormous—from 5 to 200 times the ADI.

That was in 1985; the numbers have been expanded and changed somewhat now. But the basic point has not changed at all. *If* all foods contain pesticides at the tolerance level (which they don't), and *if* the EPA's definition of safe exposure is accurate (which is questionable), a child eating an average child's diet is exposed to pesticides at well above the level defined as safe.

Does that mean that our children are being poisoned? There were too many *ifs* in the previous paragraph to know. No one knows, probably not within a factor of 100 or 1,000.

The kids might be safer than these researchers' calculations suggest because most crops contain residues well below tolerance levels, but the sampling data are too scattered to be sure of that. Some children might be more endangered than the average because they eat much more of some foods than the average child. Children are also more endangered than adults because they have poorly developed immune systems, lower abilities to detoxify foreign substances, and more permeable skins, lungs, and digestive tracts. They also have a longer lifetime for an induced cancer to develop.

Then there are all the uncertainties about multiple paths of exposure, interactions among different chemicals, and applicability of animal studies to humans, which mean that we don't know if the ADIs are accurate in the first place, for adults or children.

I've tried to use careful language here. Wargo's language is even more careful. I personally think his findings are hair-raising and should be trumpeted from the rooftops, but trumpeting is not what scientists do—and industry and government are not about to alarm the public over a possible risk with a massive uncertainty. For three years I have watched to see how the National Academy, the EPA, and Wargo himself would respond to his findings.

Wargo updated his numbers, triple-checked them, expanded them, made his program more comprehensive and easier to use. In 1987 he presented his results to the National Academy of Science and the EPA. Those results formed the foundation for the Academy's publication, "Regulating Pesticides and Foods." Thanks to the efforts of Wargo and others, Congress has authorized the Academy to conduct a $600,000 study of childhood exposure to pesticides, overseen by a committee of the nation's best pediatricians and toxicologists.

Wargo is trying to make his computer program available to environmental groups and the public. To him the issue is not only the children, but the ethics of exposing the public to any risk. He teaches a graduate course now at Yale on the science and politics of the environment, the theme of which is, as he described it to me, "If you're struggling to design an ethically responsible regulatory system, the people who are bearing the risk should have a voice in assessing that risk. There's got to be a way of getting the information to the people, getting them to understand the uncertainties. The burden of proof in an uncertain situation like this can't be on the public. That's wrong. That's just wrong."

John Wargo is also a parent. His personal reaction to his findings has been to scour southern Connecticut, where he now lives, looking for organic apple juice for his two-year-old son.

Can We Feed the World Without Pesticides?

W E MUST feed ourselves. To do that, we must have agricultural chemicals. Without them, the world population will starve," says Norman Borlaug, who won the Nobel Peace Prize for breeding high-yielding grains.

But thousands of modern, high-yielding farms use no agricultural chemicals at all.

The four Lundberg brothers of Chico, California, have a 2,000-acre organic rice farm. Some of their fields have not seen a pesticide for fifteen years.

Del Ackerlund farms 760 acres of corn, oats, alfalfa, and soybeans near Valley, Nebraska. He hasn't used commercial fertilizers or pesticides in nineteen years.

The Pavich brothers grow 640 acres of table grapes in Delano, California, and 900 acres of grapes, cotton, and melons near Phoenix, Arizona—all without chemicals.

Ben Brubaker's 300-acre corn, hay, small-grain, and cattle operation in Kutztown, Pennsylvania, has been organically farmed for sixteen years.

All these farmers get high yields. All of them are prospering.

How do they do it? By using every technique they know, except the spraying of pesticides.

Relay cropping, for example, controls weeds without herbicide. Last August I stood in a weed-free, relay-cropped soybean field in Pennsylvania. In the spring the soybeans had been drilled right into a greening-up crop of winter barley. The barley ripened and was combined off in July, leaving the soybeans to make a fall crop. Two harvests a year, with the crops so dense there was never enough space for a weed to get going.

Other weed control methods include plain old tilling, ridge tilling, mowing, mulching, cover cropping, flaming, and, above all, crop rotation. Planting the same crop at the same time year after year allows a buildup of weeds. Switching crops and planting times keeps any single weed from getting established.

Crop rotation helps control insects too. So does good soil. Just as healthy people are less likely to catch colds, healthy plants are less likely to host bugs. Healthy plants come from soil full of humus and microbes, many of which attack plant pathogens.

Organic farmers are fanatic about their soil. The Lundbergs refuse to burn rice straw as their neighbors do; they return it to the soil. Craig Gardner, an organic apple grower in Sebastopol California, is doubling the humus content in his orchard with composted apple pomace and manure. Gregg Young, an agricultural consultant in California, has found that high soil calcium levels help control nematodes, fire blight, and grape bunchrot.

Organic farmers encourage natural predators: ladybugs to eat aphids, brachonid wasps to parasitize tomato hornworms, *bacillus thuringiensis* to infect cabbage worms. They use traps, such as sticky red balls in orchard trees to attract the apple maggot fly. They inter-plant repellants, such as catnip or tansy herbs, around squash plants to repel cucumber beetles. They plant resistant varieties.

They know what they're doing. Knowledge is the most important input to modern organic farming—knowledge of how to harness free natural processes instead of expensive chemical ones.

Lack of knowledge is the main reason why organic agriculture is practiced on only about 1 percent of the farms of North America and Europe. Most of the sources of farmers' information—journals, agri-culture schools, extension services, advertisements—push chemicals. Monsanto can sell an herbicide and then a genetically engineered soy-bean resistant to that herbicide. It can't sell the idea of relay cropping.

Only recently has the knowledge situation changed. Several land grant universities, such as Nebraska, Michigan, and Pennsylvania, are doing serious research on organic methods. The California Legislature has appropriated special funds to the University of California for that purpose. Rodale Press puts out a journal on low-input agriculture called *New Farm*.

The International Federation of Organic Agriculture Movements (IFOAM) allows worldwide exchange of information on organic prac-tices. IFOAM's international conference in August 1986 was attended by 400 people, from every continent except Antarctica. I was one of them.

I listened to presentations on organic farming, small-scale and large, in warm climates and cold. I was impressed by how far this technology has come, with so little help from governments, universities, or indus-try. The basic crops, most grains, and many vegetables give high yields with no pesticides now. For some special crops, fruits, and some tropical situations, there is a great need for more research and experi-mentation.

The world's farmers pay $12.8 billion per year for pesticides. That is the direct cost. The indirect costs are contaminated groundwater, disrupted ecosystems, industrial accidents such as Bhopal, hazardous waste dumps, poisoned farmworkers, costly regulatory systems, and increasingly resistant pests.

Those costs are not the price we have to pay to feed the world. They are the price of clumsy technology, blasting the environment with poisons, rather than working knowledgeably with nature's methods of controlling pest populations.

The Dilemma of Dursban and Dartmouth's Elms

SKELETONS OF elm trees still stand in the valley where I live. They have succumbed to Dutch elm disease so recently that they haven't yet been cut down. The dead branches hold their curving V shapes, like huge elegant vases, reminders of a beauty that is now only a memory.

Only a memory, except on the Dartmouth College campus, where 150 American elms are still alive and graceful, casting welcome summer shade on the lawns. About half of them are mature trees, their branches arching 50 to 80 feet overhead. The other half are youngsters. Bob Thebodo is actually optimistic enough to plant new elms.

Thebodo is in charge of Dartmouth's 1,200 campus trees. He has every one of them mapped, has a computer printout with each tree's vital statistics, knows each personally. "That big elm back of Baker Library, that's my favorite. Hasn't given me a bit of trouble."

Though he likes all trees, Thebodo has a special respect for elms. They're tough, he says, a good urban tree, tolerant of salt, pavement, and pollution. They're only bothered by Dutch elm disease and construction. Every time Dartmouth digs up a phone line, a sewer line, or a steam pipe, the roots of an elm are likely to get cut. The weakened tree attracts the elm bark beetle, the carrier of the fungus that causes Dutch elm disease.

Elms may be tough, but Thebodo has to give them a lot of help these days. Sanitation is the first line of defense; Thebodo cuts back immediately any limb that shows a sign of infection. The second defense is fertilization, keeping the trees healthy enough to resist disease.

The third line of defense is chemotherapy—injecting into the vascular system of each tree a fungicide called Arbotect. The earth is carefully dug away to expose the root flare several feet down. Holes drilled into the roots are attached to tubes through which flows dilute

Arbotect ("$215 a gallon—liquid gold"). One treatment is good for three years; one-third of the trees must be done each year.

The fungicide and the drilling do some damage to the tree's tissues. Thebodo doesn't know how long this treatment can be continued before it does more harm than good. He hopes he can keep it up indefinitely because the chemical directly attacks the fungus, which cannot be said for the fourth and most controversial line of defense.

Every spring Thebodo sprays the insecticide Dursban to knock back the population of elm bark beetle. It isn't easy to spray a full-size elm tree. You have to suit up in a rubber outfit, gloves, and a respirator and either shoot down from a cherry picker or up with a high-powered sprayer. Spraying can be done only on rainless, windless nights when the temperature is above 40 degrees. Back when he sprayed more trees more often, Thebodo had to wait many nights for the right weather, then work all night to get the job done.

Before Thebodo's time the insecticide of choice was DDT and the spraying was frequent. When DDT was banned, Dartmouth turned to methoxychlor, twice a year on all elms. Now Thebodo prefers Dursban, one of the most commonly used horticultural chemicals in America.

The active ingredient in Dursban is chlorpyrifos, an organophosphate. It affects the membranes of nerve cells, disrupting the pattern of nerve firing. Therefore, it deranges virtually every body function, not only in insects but in any creature that has nerves.

Dursban is toxic to birds, fish, crustaceans, and bees. In human beings it causes headache, confusion, convulsion, blurred vision, vomiting, fatigue, and tremor. Dursban has not been tested for its tendency to cause cancer. In fact, it may not have been properly tested for anything since its federal approval was based on data provided by Industrial Bio Test Labs, whose executives have been sent to jail for faking test results.

Every time Bob Thebodo sprays Dursban, he gets angry phone calls from a few people in the community. Thebodo sympathizes with them, but he says he has done everything he can think of to reduce his use of pesticides—everything except give up the battle against Dutch elm disease.

He tried pheromones, hormones that attract insects into traps. He found they were effective only against European bark beetle; the beetle infesting the campus is a native one. He tried girdling a few trap trees, unwanted young elms, weakening them to attract beetles, so he could selectively spray only those trees. That didn't slow down the disease. He planted "disease-resistant" varieties, and they promptly contracted the disease.

Until he thinks of something else to try, Thebodo has reduced the Dursban spraying to only the biggest, most scenic, most stressed trees around the central Dartmouth green, and he just endures the hostile phone calls.

"I try to be in the middle, not spraying every bug, but not avoiding all spray either. These people are yelling at me about something I don't like to do."

With all his precautions, Thebodo loses five or so elm trees a year. He mourns every one that comes down, and so do I. I drive past the elm skeletons along the roadsides and then come to Dartmouth's living trees and rejoice. That's difficult to admit because I also happen to be a pesticide hater. I thoroughly distrust the federal pesticide regulatory system. I agonize when poisons are sprayed where people, bees, song-birds, and earthworms have to live. I refuse to use sprays on my farm, and I've lost every elm tree.

So here I am, a bundle of contradictions, hoping Bob Thebodo can keep those magnificent trees alive long enough and experiment persistently enough to find a way of protecting them without having to drench the land with poison.

Allergic to the Twentieth Century

UNTIL SHE was twenty-six, Annie Berthold–Bond had never had a medical problem. That year her husband Daniel was a graduate student at Yale, and she had just finished art school and was working as a waitress. One day there was a gas leak at work. "Lots of people passed out. I didn't, but now I see that was a turning point."

Annie was tired and depressed all that winter, as if she were getting a flu. Doctors found nothing wrong. She and Daniel spent the following summer in Hanover, New Hampshire, where their families lived. Outdoors and active, Annie felt much better.

In the fall she walked back into their New Haven apartment and smelled something strange. An exterminator had treated the place with pesticide. Within three weeks she had to quit her waitressing job. "My body was rigid with tension, and I had crashing headaches. I couldn't concentrate, I couldn't read or watch television. I had no strength. I'd wake up at 4 A.M. obsessed with worry."

The Yale doctor sent her to the mental health clinic, which put her into therapy for an emotional problem.

In January Annie's apartment building was treated with pesticide again. She went into an emotional nosedive. "I couldn't believe the violent thoughts I was having. I was afraid I was going to kill someone."

As her aggressive mood grew worse, Annie convinced Daniel to take her to the hospital. She stayed two nights, medicated on Valium, then was sent home. The problem came back. Desperate, Annie committed herself to the hospital again. She was diagnosed as an "atypical manic depressive" and consigned to the psychiatric ward. She stayed in the hospital two months on antidepressants.

Annie might have been in and out of mental hospitals for a long time if it hadn't been for her sister Kathy, a biochemist and the mother of a hypersensitive child. Kathy had concluded that her son had allergic reactions to junk foods and petrochemicals. She suggested that Annie try some allergy tests.

That was another turning point. Annie's response to common allergens tested off the scale. Her immune system was wildly overreactive. She stopped looking inside her mind for the source of her trouble and started focusing on her environment.

She quit painting with oil-based paints. She tried to stay away from pesticides. But she and Daniel had just moved to Bard College in upstate New York, and the fresh paint, sheetrock, and joint compound in their newly renovated house made her sick. Then Bard sprayed the campus with the herbicide 2,4-D to kill dandelions. Annie's aggressive symptoms started up again. Pesticide drifts from an orchard three miles away made them worse.

Gradually, though, Annie found help. She joined HEAL, the Human Ecology Action League, a group of about 5,000 chemically sensitive people. A HEAL member told her about Dr. Neil Solomon of Baltimore, one of a small number of doctors who treat her ailment, which is variously called multiple chemical sensitivity (MCS), environmental illness, total allergy syndrome, chemical susceptibility, or profound sensitivity syndrome.

Dr. Solomon helped stabilize Annie's immune system and counseled her on how to avoid triggering chemicals. She recovered slowly, except for maddening setbacks whenever she ran into any of the thousands of traps the modern world sets for the chemically sensitive.

Sunlight striking a bulletin board and vaporizing its binding chemicals could make her sick. She reacted to residual pesticides in melons. She couldn't ride in a car with a badly tuned engine. She had to remove the swab soaked with fungicide that is in the receiver of every telephone. During her summers in Hanover, whenever Dartmouth College sprayed its elm trees, she had to stay out of town for a month. She and Daniel had to move again and again in search of a chemically clean home.

They can't live within five miles of an orchard or near agricultural land, power lines, train tracks, golf courses, smokestacks, garbage incinerators, gas stations, toxic dumps, major highways, or many kinds of businesses. They can't use synthetic fibers—only cotton or wool untreated with stain protectors, pesticides, or flame retardants. No fresh paints, stains, varnishes, urethane. No cosmetics. They clean with baking soda and unperfumed soap. They plead with neighbors not to spray fruit trees or roses—and the neighbors are not always understanding.

You could say that Annie is allergic to the twentieth century. She's not the only one. The Environmental Health Center in Dallas treated 17,000 MCS patients between 1976 and 1988. Canada recognizes MCS and offers many social services to its victims. The Board on Environmental Studies and Toxicology of the National Academy of Sciences estimates that 15 percent of the U.S. population is sensitive to common household chemicals.

Neil Solomon guesses the percentage is even higher. He thinks that everyone has a threshhold of exposure, beyond which hypersensitivity can develop. Some people, like Annie, have lower threshholds than others. They are canaries for us all, able to sense poisons that we can't detect.

How many people with chemical sensitivities are in mental institutions or on drugs that can't help them? How many children are being punished for behavioral problems, when in fact they're having allergic reactions? No one knows.

What are the rights of these sensitized people? No one has thought it through. Annie says, "I try to get the neighbors to stop burning trash or the town to stop spraying trees, and I'm treated as a nuisance. It's so demoralizing."

Annie and Daniel are slowly learning how to exist in a petrochemical world. Annie is starting a service called Environmentally Safe to help others learn too. She is especially interested in helping pregnant women because in 1988 she finally felt well enough to have a baby. Her daughter Lily is smiling, roly-poly, and healthy.

"There's something about having Lily now—I don't feel like such an aberration any more. Those years were devastating to my self-esteem. Now I know I have to fight these chemicals." Annie told me her story calmly, in precise detail, in her carefully planned, chemical-free home. But the years of fear and frustration have not left her entirely. As I was about to leave, she stopped me and asked, tentatively, a bouncing Lily in her arms, "You do believe me, don't you?"

Home, Home on the Loam

IT'S EARTH Day 1987 at the University of Wisconsin. A retired professor of soil science, with the appropriate name of Francis D. Hole, has been asked to address the Environmental Conservation class. He is a white-haired pixie, who exudes more energy than all 120 laid-back students combined. He is carrying a fiddle.

"How old are your feet?" he begins. "Well yours are about twenty years old and mine are seventy-three. But actually our feet are millions of years old, and they're attuned to the earth. When they're on a hard floor like this, they think they're in a desert. Hardpan. Not much water, not much food. Ugh! But when they walk on the nice, springy grass, they know they're in touch with *life*!"

Professor Hole quotes a snatch of Walt Whitman:

> Underfoot the divine soil
> Overhead the sun.
> The press of my foot to the earth
> Springs a hundred affections.

"Notice the *bounce* of the life-giving soil! It can be a *wonderful experience* to walk over to the chemistry building," he says, bouncing. "When you're walking on good soil, *all nature* is singing with you." On the fiddle he plays the sound of trudging footsteps and then adds a cheerful obligato, nature's song.

He's not the world's best fiddler, but he's the most unapologetic. He's glowing with enthusiasm. The students are trying to look cool.

Francis Hole produces with his fiddle the sound of roots growing down into the soil. He adds a high vibrato. That's the little soil bugs, he says, the mites and bacteria, the fungi and algae, throbbing away, producing nutrients for the roots. "It shouldn't be called terra firma at all; it should be called terra *vibrata*!"

He launches into a soil song, pointing out that anyone who thinks soil is just plain brown has not been paying attention. He sings:

> A rainbow of soils under our feet,
> Red as a barn and black as peat.
> Yellow as lemon and white as the snow,
> Bluish gray are the colors below.
> Hidden in darkness as thick as the night,
> The only rainbow that forms without light.

On to a tune everyone knows. By now the students are softened up enough to join in.

Oh give me a home on a deep mellow loam
That supports the trees and the grass;
Where we hardly recall a bad crop year at all,
And the crickets rejoice as we pass.

"You can't have a range without a loam," says Professor Hole. "We ought to make our songs about what's *really* important!

"We have everything backward. Do you know it took Wisconsin until 1983 to adopt an official state soil? Think what a pickle the badger, the state animal, was in without a state soil to burrow in. What do you think has been feeding the state tree, the maple? We should have had a state soil *first*, not last!"

Wisconsin's state soil is the Antigo Silt Loam, found in twelve counties in the north central part of the state. It's not the best soil in Wisconsin, nor the worst. It grows potatoes and forests and pastures for America's dairyland. Francis Hole, of course, was the motor behind the seven-year effort that led to its adoption. He had a lot of help from schoolchildren, who bombarded legislators with letters. Most of them were inspired by Francis Hole's puppet show, in which Bucky Badger helps save Antigo Loam from the nasty monster Erosion.

On the day Antigo Silt Loam became Wisconsin's official state soil, Francis Hole stood with his fiddle on the steps of the capitol building and sang the Antigo Silt Loam song, which he now teaches to the Earth Day class in Environmental Conservation.

Antigo, a soil to know,
Wisconsin's crops and livestock grow,
and forests, too, on Antigo,
and forests, too, on Antigo.

Great Lakes region, fertile land;
Glaciers spread both clay and sand;
Winds blew silt, then forests grew,
Giving soils their brownish hue.

"Wow, a whole course in soil science, right there in *one verse*!"

Great Lakes region, fertile land,
You strengthen us in heart and hand;

> Each slope, each flower, each wild bird call
> Proclaims the unity in all.

"Now that's what I'd call *awareness!*"

There's time for a few more songs—"Where have all the bedrocks gone? (Long time weathering)" and "You are my soil, my only soil, you keep me vital, night and day." By now the unabashed joyfulness of Francis Hole has infected everyone in the room. The students are singing and whistling and stamping.

The class period ends. Everyone goes off to chemistry on million-year-old feet, bouncing on the grass that grows out of the loam, listening as all nature sings.

ENERGY

WE CAN

LIVE WITH

Just as I am convinced that we can feed the world with organic farming, I am also convinced that we can provide all the energy we need without air pollution, oil spills, or a greenhouse effect. We can do it with renewable solar sources, used with high efficiency. It will take a while to build up that new energy system, but it will require much less cost and effort than trying to proceed with either fossil fuels or nuclear power.

Some people would like environmentalists to come around to the position that we could also create a sustainable world with nuclear power. Environmentalists are not likely to do that until they are convinced that nuclear wastes can be handled safely for the next 100,000 years—and that is an argument that no properly humble human being can make.

If the alert reader notes a certain testiness toward nuclear power in the columns here, it comes from long and intimate experience with the nuclear industry. As a student I worked at the Argonne National Laboratory near Chicago, one of the homes of the reactor and the bomb. I've been fighting nuclear power ever since. On most hotly argued social issues I can see both sides, or at least I'm willing to try. On the question of nuclear power I make no apologies for dedicated one-sidedness. Environmentally, economically, and ethically, I think there's no justification for running a single nuclear reactor, for purposes of either war or peace.

123

The Ups and Downs of the World Oil Market

[NOTE: The following column was written in 1985. I have not edited it to include mention of the sudden changes in the oil market that took place in the summer of 1990. The very contrast between this view from 1985 and the view five years later makes my point.]

Some of my neighbors are selling their woodstoves. Others are buying bigger cars. Makes sense. Gas is getting cheaper, right?

The government is undoing the oil conservation measures installed during the petroleum panics of the 1970s. Automobile fuel economy standards are being rolled back. Solar research is essentially gone. The synfuel program is dead. The energy crisis is over.

Yet if there's anything sure about the world's oil supply, it's that there is much less of it now than there was ten years ago. We burn about 20 billion barrels a year, and the planet doesn't make the stuff anywhere near that fast.

How can it be that oil is more scarce and yet cheaper? To understand that you have to understand the difference between geology and economics.

Geologists tell us that the discovery rate for new oil has fallen behind the consumption rate for fifteen years now. Known reserves are slowly falling. In terms of real capital, energy, and effort, oil is more costly (though not higher priced) than it used to be, and it will get ever more so as we take it from deeper down and farther afield. The U.S. Geological Survey says that substantial production of oil will be over within fifty years. Natural gas will last a decade or two longer. Children born today will live to see a world no longer powered by oil or gas.

Economics deals not with real costs but with prices. Prices are social inventions, subject to the pull and tug of competitive and political forces. They are only loosely related to oil in the ground.

In the short term, oil price reflects oil pumped, processed, and ready to deliver relative to the unfilled orders for that oil. This is the spot market. It can change radically with any event that interrupts the flow of oil or that makes speculators *think* the flow will be interrupted. The spot oil price gyrates when the Strait of Hormuz is threatened, or when there's an unusually cold winter, or when the Ayatollah sneezes.

Underlying the short-term variations in spot price are the decades-long swings of the medium-term oil price. This price measures not pumped oil versus immediate demand but oil-pumping capacity versus oil-burning capacity. It measures the extent to which operating oil wells can supply operating furnaces, cars, and boilers.

During the early 1970s oil wells were working at over 90 percent capacity. Short-term shocks simply could not be met by pumping harder, so prices rose instead. Gradually, in response to the higher price, more oil wells were drilled, and meanwhile we were all conserving. Supply capacity rose, demand capacity leveled off. It didn't happen fast because it takes a while to locate and drill oil wells and to turn over the car fleet and insulate houses.

During that delay, we overshot. By 1985 we reached an overbalance of oil wells to oil burners. Middle East wells were operating at less than 50 percent capacity. OPEC was weakening, and the price went down. Few new wells were being drilled. Consumers were buying bigger cars. As demand rises and supply levels off, we are setting up the next swing—the oil scarcity of the 1990s.

In other words, we have a glut right now not of oil but of oil wells.

The long-term price measures the real cost of finding, producing, and transporting oil. At the moment it is almost buried under the short- and medium-term noise. But real cost is going up faster and faster as the closest, easiest-to-find deposits get used up. Some day this cost will dominate the short- and medium-term forces, and we will finally have economic evidence for what the geologists have always known—the earth's petroleum deposits are finite and non-renewable.

The price of oil is a useless guide to energy policy. It swings us around on its cycles of over- and underinvestment in oil wells. "For decades oil prices went down, and we thought they would always go

Figure 4

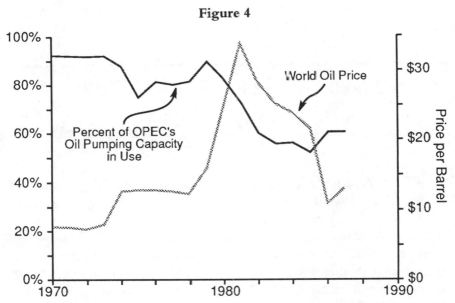

Capacity utilization of OPEC's oil wells, versus world oil price. The two variables stand in both cause and effect relationship to each other. High capacity utilization makes it possible for price to rise. High price brings on more capacity (in OPEC and elsewhere), which then makes price fall.

SOURCES: US Department of Energy; Oil & Gas Journal

down," says Dr. Ulf Lantzke, director of the International Energy Agency. "Then in the 1970s they went up, and we thought they would always go up. Now they are going down again, and we think they will always go down."

Over the next fifty years we will have to replace every oil-burning device in the world with one that uses some more sustainable form of energy. It's a technical challenge equivalent to the one that moved us from horses and coal to oil and gas at the beginning of this century. The less we waste oil, the longer we will have for this transition.

It may not make medium-term economic sense, but it makes geological sense to tighten fuel-economy standards and speed limits and to encourage better insulation, cogeneration, and all the other conservation options. It makes sense to develop solar energy. It even makes sense to keep that high-mileage car and that woodstove.

Alaska Oil Versus Alaska Wildlife:
Many Voices

T HE NORTH Slope is a flat, crummy place. Only for oil would anyone want to go there," says an official of the Arco Oil & Gas Company.

The May 1988 issue of *Audubon* shows picture after glorious picture from that "flat, crummy place"—musk-ox, tundra swan, Arctic fox, snowy owl, brilliant wildflowers, braided rivers, and magnificent mountains—all in the Arctic National Wildlife Refuge on the North Slope of Alaska.

The proposition is that this Wildlife Refuge should be opened to oil exploration. The debate is heating up. Voices are rising. Claims are contradictory. Either the opportunity to make the last major on-shore oil discovery in the United States is about to be lost or a pristine wilderness is about to be lost, depending on who's talking.

The American Petroleum Institute: "The petroleum industry cares about the special ecological and wildlife values of the Arctic and works hard and spends great sums to protect them. The record at Prudhoe Bay proves we have been good stewards of the land."

A slick, colorful publication of the Exxon Corporation: "North Slope oil is now providing about one-fourth of all domestic oil production with no significant impact on the ecologically sensitive Alaskan environment."

An equally slick, colorful publication of the National Audubon Society shows black smoke pouring into the sky at Prudhoe Bay, bulldozers, drilling rigs, and garbage dumps. There have been 23,000 oil spills there. A report to Congress by the Fish and Wildlife Service says that oil activities have stripped 11,000 acres of Arctic vegetation and have reduced the numbers of most bird species.

John F. Seiberling, chair of the Alaska Lands Subcommittee of the

Congress: "Twenty of twenty-one major waste-storage pits at Prudhoe were found . . . to violate Environmental Protection Agency standards, discharging toxic chemicals, heavy metals, and carcinogens
into wetland habitats."

Vice President Bush: "I'd like to see us open up that Alaska refuge. It
was said once, remember, when they built the pipeline, 'Don't build
the pipeline, you get rid of the caribou.' The caribou love it. They rub
up against it and have babies. There are more caribou in Alaska than
you can shake a stick at."

The caribou population has increased around Prudhoe Bay, but not,
says the Fish and Wildlife Service, because of the aphrodisiac effect of
pipelines. The roads constructed by the oil industry have opened the
area to hunters, who have decimated the populations of bear and wolf,
predators of the caribou.

Oliver Leavitt of Barrow, Alaska, vice president of the Arctic Slope
Regional Corporation: "Environmentalists don't give you hospitals,
roads, housing, water, and sewers."

The mayor of the North Slope village of Kaktovik: "You can't move
the oil somewhere else. It's there, and this country is going to get it
sooner or later. I believe we can get it in a way that will not cause
significant damage."

According to *Audubon* the "full-leasing" scenario would bring to the
Refuge four airfields, 300 miles of road, 100 miles of pipeline, two
desalinization plants, seven large production facilities, fifty to sixty
drilling pads, ten to fifteen gravel pits, and a seaport or two.

L. G. Rawl, chair of the Exxon Corporation: "The entire nation will
forfeit potentially substantial economic benefits if oil exploration of
this narrow strip of tundra—about one-half of 1 percent of Alaska—is
unduly delayed."

Amory Lovins of the Rocky Mountain Institute in Old Snowmass,
Colorado: "Leasing in the Arctic National Wildlife Refuge has an 81
percent chance of finding no economically recoverable oil; a 19 percent
chance of finding oil averaging a six-month national supply; a 1 percent chance of a year and a half's worth; and a 100 percent chance of
trashing the refuge. If such poor odds of so little oil are 'vital to our
national security,' why cut new-car standards from 27.5 to 26 miles a
gallon—thus wasting more oil per year, with 100 percent certainty,
than unlikely success in the Refuge could yield?"

An Anchorage bumper sticker: "Please, God, send just one more oil boom. I promise I won't blow it next time!"

Former New Mexico Senator Clinton Anderson: "Wilderness is an anchor to windward. Knowing it is there, we can also know that we are still a rich nation, tending to our resources as we should—not a people in despair searching every last nook and cranny of our land for a board of lumber, a barrel of oil, a blade of grass, or a tank of water."

A Negabarrel Saved Is a Negabarrel Earned

AMORY LOVINS shuttles around the world carrying a small suitcase that contains, he says, 200 large power plants plus all the oil that runs through the trans-Alaska pipeline.

What it really contains is energy-saving lightbulbs, motors, showerheads, and other devices. They help Amory get across the most basic points about energy, namely:

1. No one *wants* energy. No one is pining to have a kilowatt-hour of electricity or a barrel of oil for its own sake. What people want are energy *services*: hot showers, cold beer, the ability to travel at reasonable speed and in relative comfort.

2. Energy is not a benefit but a *cost* of getting energy services. The less energy we have to spend to heat the shower or cool the beer, the better off we are.

These are simple enough points, but the policy of nearly every country ignores them. Nations pride themselves on energy production, not energy efficiency. They search the wilderness for oil, pump it from the bowels of the earth, fight wars over it, invest billions in electricity plants. Oil companies and electric utilities think of themselves as selling energy rather than as selling energy services at the lowest possible cost.

Amory Lovins and his colleagues at Rocky Mountain Institute are trying to get across the idea that more efficiency can produce the same results as more energy for less money and with less environmental damage. If we can turn a motor, light a bulb, or cool a room with fewer watts or barrels, that's just as good as discovering new barrels—better, actually, because a discovered barrel can be burned only once, while an insulated house or more efficient car goes on saving barrels over its whole lifetime.

As Lovins puts it, "In the lower 49 states are two supergiant oil-fields, each bigger than the largest in Saudi Arabia; each able to produce sustainably (not just to extract once) almost 5 million barrels per day for less than seven dollars a barrel; each capable of *eliminating* U.S. oil imports before a new synfuel plant or power plant or frontier oilfield could produce any energy whatever, and at about a tenth of its cost. They are the 'accelerated-scrappage-of-gas-guzzlers oil-field' under Detroit and the 'weatherization oilfield' in the nation's attics."

Lovins carries conversion factors in his head and a calculator in his pocket and writes in dense sentences, amply footnoted. Those sentences contain some amazing figures. For example, "Superefficient lights, motors, appliances, and building components can together, if fully used in existing U.S. buildings and industries, *save about three-fourths of all electricity now used*, at an average cost far below that of operating a typical coal-fired or nuclear power plant, even if building it costs nothing."

Let that one sink in slowly. We could write off 75 percent of our existing power plants, invest in efficiency technologies, still have our cold beer and hot showers, and save money. And produce no nuclear wastes, much less acid rain, much less air pollution.

Maybe it's the enormity of that claim that makes it so hard to grasp. It's not easy to admit that we are constructing an energy system that violates a basic law of the free market—invest first in the cheapest option with the highest payback.

Energy efficiency is the cheapest option. Some smart investors are recognizing that. Despite national policies that subsidize supply rather than efficiency, in 1986 the United States used 38 percent less oil to produce a dollar of GNP than it did in 1973. The 1973–86 improvement in the car fleet mileage (from 13.3 to 18.3 miles per gallon) saved in 1986 more than twice as much oil as we imported from the Persian Gulf that year. From 1979 to 1986 the United States got fifteen times as much new energy from increased efficiency as from increases in nuclear plants and fossil fuels.

And we have just begun to tap the possibilities for efficiency, says Lovins, reaching into his suitcase. He pulls out a lightbulb, screws it into a socket, and as it glows, he says, "This bulb gives as much light as those 75-watt floods up there in the ceiling, but it uses only 8.5 watts and lasts four times as long."

He holds up a sample of glass coated with heat mirror—it is as transparent as any window, but it retains heat just as well as an insulated wall does. He has water-saving showerheads, electronic ballasts that save 50 to 90 percent of the power needed for fluorescent bulbs, and a device that reduces friction loss in motor gearboxes. Not in the suitcase but in his mental bag of energy-saving technologies is a Toyota prototype car that gets 98 miles per gallon and a Renault Vesta four-passenger car that road tests at 121 mpg.

Companies and localities with eyes on the bottom line are beginning to see that what Lovins calls a "negawatt" or a "negabarrel" is as good as a watt or barrel. Southern California Edison gave away over 500,000 energy-efficient lightbulbs to its customers and saved more from reduced generation than it paid for the bulbs. The Central Maine Power Company offers factories cash grants to adopt more efficient equipment. The city of Austin, Texas, has, with Lovins's help, figured out how to save seven to nine times its share of the South Texas Nuclear Project, pay its debt for that project, and lower electric rates.

Still, efficiency is a hard idea for the production-minded to grasp. A managing director of a large oil company, struggling with the upside-down thought of selling negabarrels, once remarked to Lovins that once you sell a negabarrel, you can't sell it again. Lovins replied that once you sell a barrel of oil, you can't sell *it* again either. The question should be, he said, which sale will produce the greatest energy service at the lowest cost and with the least regret?

When We're Ready for Fusion Energy, It's Ready for Us

IF THE chemists at the University of Utah have indeed achieved nuclear fusion in a bottle on a tabletop, their discovery could open a new age in human history.

The economy could run on deuterium, a component of seawater that is vastly abundant. It is available to poor countries as well as rich ones. We could air-condition our homes, microwave our food, drive our brontomobiles without producing smog or acid rain or greenhouse gases. There would be no spills from pipelines or tankers. We would not have to please the Saudis or give tax breaks to oil companies or defend the Persian Gulf.

Fusion would produce less harmful radioactive waste than nuclear fission and would not depend on a chain reaction that could go out of control. And if fusion can really happen in a lab with a beaker and some palladium electrodes, it won't need centralized, billion-dollar megaplants, manned by specially trained technicians surrounded by security systems. There would be no need for a scientific priesthood to plan our energy future for us. "Cold" fusion could make energy not only cheap, clean, and abundant but even democratic.

Who wouldn't be in favor of a breakthrough like that?

A few people have stated emphatically that they wouldn't. They are mostly ecologists, just the folks you'd think would be delighted to find a less polluting form of energy. But they are also people who have thought deeply about the role of energy in human affairs. Their conclusion was summed up by Stanford biologist Paul Ehrlich. If we were to tap cheap, inexhaustible energy, he said, it would be "like giving a machine gun to an idiot child."

To imagine what might happen if all energy constraints were lifted, you need only look at the claims people are making for the possibilities of tabletop fusion.

We would never run out of materials, they say, because we could grind up ordinary rock for its copper or palladium—forgetting that over 99 percent of ordinary rock is ordinary rock and would be left behind in huge, leaching piles. We could produce fertilizer for all the world's farmers, they say—forgetting that fertilizers run off to pollute waters. We could manufacture anything we want—with all the accompanying toxic and solid wastes.

But we could use some of our abundant energy to combat pollution, they say—unaware that most "pollution control" methods simply move waste from one place to another, from air to sludge, for instance, or from Philadelphia to Africa.

If material desires remain unlimited and our planetary consciousness remains primitive, we would just use unlimited energy to generate unlimited waste. With no energy constraint, we would run into environmental constraints all the faster.

Then there's what we'd do to each other. The scramble for tabletop fusion reveals the mentality with which the human race would use the boon of a new energy source.

The researchers have not released their precise experimental conditions nor permitted others to visit their lab. Normal scientific rules of objectivity and civility have broken down as information is conveyed through press conferences. Chemists and physicists are scrapping like four-year-olds. The University of Utah has applied for five patents. Over 200 major corporations have sent inquiries about buying their way into a piece of the action. The price of palladium has soared. The researchers were summoned to a packed congressional hearing. The government saw a way for the nation to achieve not only economic ascendancy but military ascendancy. After all, the one use to which humanity has put fusion energy so far is thermonuclear bombs.

In short, the primary human reaction to this potential boon is to try to corner it for profit or power.

I'm one of those who hope that tabletop fusion is a mirage—which it probably is. I hope that we human beings have more time to bring ourselves to live within limits, to live in harmony with one another and the earth, and to find purposes more worthy than the accumulation of power or wealth. The funny thing is, if we ever do become willing to live gently, moderately, and unselfishly, we will discover that we already have cold fusion at our fingertips.

Fusion is the power that lights the stars, including our sun. Cold fusion power falls on our heads in quantities thousands of times greater than we need, generated from a reactor located a safe 93 million miles away, with an expected lifetime of several billion years and with no need of investment or maintenance from us. We don't think much of this power because it comes at a regulated pace that can't be hustled. No one can corner it. By the crazy reckoning of our economics, which counts only benefits to some human beings and discounts costs to most human beings and to the environment, we think it's expensive. But it's not expensive, it's not polluting, and it's there for us to use as soon as we are wise enough to use it.

The scramble for tabletop fusion reminds me of the ancient Zen story of the seeker going all over the world searching frantically for an ox while riding on the ox. That's a metaphor for the search for enlightenment, which is in fact as close at hand as one's willingness to transcend one's own egotistical desires. It could also be a metaphor for the search for energy—which is just as nearby as enlightenment, and just as available on exactly the same terms.

Nuclear Waste in Everyone's Backyard

I T SEEMS that New Hampshire is blessed with the sort of deep granite the Department of Energy has been looking for, in order to create the nation's second high-level nuclear waste depository.

The first depository will be out West—the choice there has been narrowed to three sites in Texas, Nevada, or Washington. The second site is scheduled to be chosen in 1991 from among twelve contenders, one of which is about thirty miles from my farm.

Someone once said that the good thing about being under a death sentence is that it focuses the attention mightily. You can say the same things about a nuclear dump coming to your neighborhood. Nuclear power is no longer a distant concept. You begin to pay very close attention.

Representatives from the Department of Energy (DOE) have been touring the region, informing us about their beautiful plans for secreting away nuclear waste. My neighbors have been asking them some pretty good questions. Among them are these:

- Why should we believe you when you say this depository will be safe, when on the subject of nuclear power you have lied and lied and lied?
- If this dump will be so safe, why are you only considering putting it in rural areas?
- Since we know so little about storing this waste, why not keep it above ground where we can keep an eye on it?

At first I just thought these questions were funny. As intended, they made the DOE officials squirm. But the more I think about them, the more serious they get.

The government has indeed lied repeatedly to the public about nuclear power, starting in 1945 when it denied that those mysterious flashes from Alamagordo could have anything to do with the dead sheep downwind. We were told that fallout from atmospheric testing was not really dangerous, that weapons-grade material could never disappear from the government's tight surveillance, that nuclear power would be too cheap to meter. We were told only belatedly about leaking drums at Hanford, Washington, about the radiation exposure of servicemen at Bikini Atoll, about the breeder meltdown near Detroit. We heard so many obfuscations during Three-Mile Island that most of us lost whatever faith we still had in official statements about nuclear power, especially reassuring ones.

The government can't help lying about nuclear matters. It operates according to the basic bureaucratic rules of information blurring. Use long, vague words. Never admit a mistake. When embarking on something new, pretend you are far more certain than you really are. The path of nuclear power is especially littered with lies because the technical nature of the undertaking encourages long words, necessitates many mistakes, and involves awesome levels of uncertainty.

Which is what bothers me about the waste storage site. The government is setting out to dispose of complex mixtures of hellish substances, which ceaselessly generate heat and radiation. This material must be kept away from all forms of life for a period longer than any human civilization has lasted. No one knows how to do that. The storage of nuclear wastes is a superhuman job, and governments are only too human. This stage in the development of nuclear power, like all the stages before it, will be trial and error, error, error.

Who should be exposed to the error? The government's answer is "as few people as possible, therefore we should go to the boondocks." Out here in the boondocks, we favor another answer. "Those who make the decisions should bear the risks."

Why not, indeed, leave the wastes aboveground where those who generate them can watch over them day and night? There's a five-sided space inside the Pentagon that would be just dandy for the wastes from bomb making. Wastes produced by DOE research programs would fit nicely in the courtyard of DOE's Forrestal Building, right there in downtown Washington.

The rest of us can't be left off the hook. We all use electricity. We should have at least one nuclear waste storage site in each state, receiving that state's share of waste generated by its nuclear electricity consumption. The storage facilities could be on the statehouse lawns, or the grounds of the public utility commissions, or the corporate headquarters of reactor manufacturers.

That way our exalted decision-makers could see the delivery trucks coming and going. They could watch the barrels mount up. If after twenty or fifty years a canister started leaking, someone would catch on right away. We boondock folks suspect there would be much more care taken about that in Washington, DC, than in Washington, NH (pop. 366).

Maybe, with the stuff right there in their midst, more People of Importance would finally focus their attention on nuclear power. Maybe they would wonder how we could have been generating horrendous wastes for forty years without the slightest idea what to do with them. Maybe our energy options would be reconsidered, including solar and conservation. Someone might even ask whether any use to which we might put nuclear power justifies obligating the human race to a guardianship that must last 10,000 years.

Information About Energy America Can't Count On

Y OU'VE SEEN the ad. An elongated green cartoon man holds a candle and faces imposing stacks of OPEC oil barrels. The headline reads, "Nuclear energy helps keep us from reliving a nightmare."

Ads like this show up regularly in major publications and on television. They come from the U.S. Council for Energy Awareness (USCEA)—"Information about energy America can count on."

Is it the energy or the information America can count on? Neither, it turns out.

The USCEA is a coalition of suppliers, financiers, and users of nuclear power, including Westinghouse, Bechtel, General Electric, and many utilities. Its purpose is not to inform Americans about energy options but, according to its own internal documents, "to assure that nuclear power plants now in operation are allowed to remain in operation" and "to assure that nuclear plants now under construction are allowed to go into operation."

The information put out by USCEA is decidedly slippery. Here, for example, is a close look at that ad with the little green man and the OPEC oil barrels.

The ad starts out:

"The 1973 Arab oil crisis is a haunting reminder of the darker side of foreign oil dependence. Since then, America has turned more to electricity from nuclear energy and coal to help restore our energy security."

Only 4 percent of the electricity generated in the United States in 1985 came from oil—and only 0.2 percent from Middle Eastern oil. *Electricity* is simply not vulnerable to Arab oil embargoes.

Here and throughout the ad there is a deliberate confusion between oil and electricity. Nuclear power provides only electricity. It cannot replace many uses for oil—it cannot run vehicles, at least not vehicles like the ones we have now. But watch the repeated logical slips between oil and electricity throughout this ad. There's one in the very next sentence:

"As a result, these [nuclear and coal] are now our leading sources of electricity and a strong defense against an increasing oil dependence that again threatens America's energy security."

Nuclear and coal are indeed the two leading sources of electricity, but nuclear has reached second place only recently and just barely. Here are the generation figures for 1985:

> coal: 1400 billion kilowatt-hours (57 percent of the total)
> nuclear: 384 billion kilowatt-hours (15 percent)
> natural gas: 291 billion kilowatt-hours (12 percent)
> hydropower: 281 billion kilowatt-hours (11 percent)

The ad misleadingly puts nuclear in the same class as coal.

Nuclear currently supplies just 5 percent of total U.S. primary energy, about half as much as solar, wood, and other renewable energy sources (and it receives 34 percent of federal energy subsidies).

"America imported four million barrels of oil a day in 1985. Last year that increased by another 800,000 barrels a day. Most of these new barrels come directly from OPEC."

The United States in 1986 imported about one-third of the oil it uses, mostly from Mexico and Canada. The fraction of U.S. oil coming from Arab OPEC countries rose in 1986 from 4 to 7 percent. But how did the subject come back to oil? Oil is not electricity, remember?

"U.S. Interior Secretary Donald Hodel recently warned that 'OPEC is most assuredly getting back into the driver's seat' and our increasing dependence will be 'detrimental to the country's economic and national security and its financial well-being.' "

The surest and cheapest way to keep OPEC out of the driver's seat would be to create decent mass transit systems, raise efficiency standards for automobiles, and insulate our houses. We could pay for these measures many times over with the $50 billion per year we spend trying to defend the Persian Gulf.

"America's electric utilities have helped diminish OPEC's impact. Today, over 100 nuclear plants make nuclear energy our second largest electricity source, behind coal."

Since 1978 America's electric utilities have ordered no new nuclear plants and have canceled seventy-five plants on order. They have done so because nuclear plants are expensive, hard to manage, and slow to construct and because there is still no safe way to dispose of their wastes.

"*And nuclear energy has helped cut foreign oil demand. It's saved America over two billion barrels of oil since 1973, and our nuclear plants continue to cut oil use.*"

Since 1973 energy conservation measures have saved five times more oil than has coal, and ten times more oil than has nuclear.

"*Nuclear energy and coal can't offer us guarantees against another oil crisis.*" Right.

"*But the more we hear about the return of OPEC dominance, the more we need to remember the critical role played by electricity from coal and nuclear energy in fueling America's economy and protecting our future.*"

Notice how coal sneaks into that sentence in an attempt to make it factual? And notice that USCEA never mentions in this entire diatribe about energy security the fact that 50 percent of the uranium used in nuclear power plants is imported.

USCEA spends about $17 million a year to put out information like this, full of loose wording and statistical sleight of hand. It does so as a nonprofit organization, subsidized by taxpayers. Under the guise of informing us about energy, it is selling us nuclear power.

And the nuclear industry wonders why it has trouble with public credibility.

Welcome to the Evacuation Hearings

A LOT OF things about nuclear power make me mad, but nothing makes me madder than the way its bureaucrats make ordinary citizens feel small.

Suppose you live in Seabrook, New Hampshire, and you decide to go to the hearings on the evacuation plan for the just-finished Seabrook nuclear plant. It's a matter you might have an interest in—you are a potential evacuee.

You go to the statehouse and find that you have to pass through a metal detector and allow your briefcase to be searched before you can enter the hearing room.

Having been certified free of dangerous devices, you pass into the chamber where three judges sit on high. Behind them are arrayed the thirty-two white-bound volumes of the evacuation plan.

In the front row facing the judges are officials from the Nuclear Regulatory Commission, the Federal Emergency Management Agency, the state government of New Hampshire, and the utilities that own the plant. Behind them is a row of "intervenors" who oppose the plant and believe the evacuation plan is unworkable. They represent the Commonwealth of Massachusetts and the towns in the evacuation zone.

As a citizen you sit farther back. This hearing is about the safety of you and your family. Your cooperation will be essential in case of a real evacuation. But only the first two rows have microphones. You are not allowed to speak.

If you have a criticism of the plan, you should have filed it as a contention months ago. Only cross-examination is permitted now. If you are not familiar with the thousands of pages of prefiled testimony, you will have a hard time following the discussion. This hearing is not designed to be understandable by the press or the casual observer.

Still, it's amazing how many observers there are and how intently they follow the proceedings. The people in the front row are doing their jobs; the people at the back are defending the safety of their homes and families. Many of them have struggled through the technical language and know as much about the plan as the experts. They could contribute a lot to the discussion, if they were asked.

The experts up front wrangle over how many parking places there are at Hampton Beach. The folks who live at Hampton Beach could help them out. The experts have assumed that traffic will be rerouted through an intersection where you know there's a median strip in the way. The experts testify that everyone will cooperate in an emergency. Your own common sense tells you that someone is likely to panic or screw up.

Parking places and emergency behavior are important because one of the crucial numbers in the plan is the ETE, the estimated time of evacuation. The front-row experts say you can get your family away from a deranged Seabrook in six hours at most. The second-row intervenors say you'll be lucky to do it in twelve.

It may occur to you to ask one of those dumb questions ordinary citizens ask. Is six hours okay and twelve dangerous? What's a *safe* ETE? Is there some time beyond which you will be in real trouble if you don't get out?

Quiet there in the back rows. There is *no standard* for a safe ETE. The Nuclear Regulatory Commission (NRC) has never defined any measure by which one can say that this evacuation plan does or does not adequately protect the public safety.

What's the purpose of the hearing then? Stick around and listen. It will become obvious by the bias of the rules and the jokes between the judges and the front row that there's just one purpose—to license the plant.

The NRC will ask friendly questions of witnesses for the utilities and hostile questions of witnesses for the intervenors. The chair will coach the front-row lawyers to make objections and cut off damaging testimony. Any mention of a fire or radiation release will be stifled on the grounds that such an accident is not possible. No testimony will be admitted about what population growth might do to the ETE, though in this rapidly growing seacoast region the number of evacuees could easily quadruple over the lifetime of the plant.

The judges have even disallowed a serious line of testimony about how many people might get radiation sickness during the evacuation. This is an evacuation plan, they say, not a medical plan. (What if people get so sick that they aren't able to direct traffic or to drive? Will the back row please keep quiet!)

The big question that will pop into your head is the most inadmissible of all. Why wasn't the problem of evacuation considered way back when the plant site was chosen, rather than now when the plant is complete?

Actually you know the answer. If people like you had thought about evacuation back then, if you had been drawn into the process of planning, if your opinions mattered, the plant would never have been built.

Everything about this hearing, from the metal detector at the door to the technicality of the language, is designed to intimidate you. If you're the kind of person who easily feels small, you'll get the message and go away.

Some of your neighbors have stuck it out for fifteen years of this kind of treatment. They may have started feeling small, but now they feel mad, cynical, frustrated, determined, and intensely sad. In a democracy things shouldn't work this way.

We ordinary citizens are capable of understanding and deciding on even such technical matters as nuclear power. How and when did we give up our responsibility for our energy system and our safety? How are we going to get it back?

A Nuclear Power Plant That Will Never Produce

ON THE bank of the Danube 20 miles northwest of Vienna stands a completed nuclear power plant, loaded with fuel, ready to start up. It has stood there, just so, for years while the Austrians argue about what to do with it. The most popular plan is to turn it into a museum for obsolete technology.

The plant, called Zwentendorf, was intended to be the first of six Austrian nuclear plants. It was begun in 1970 and completed in 1978 at a cost of 8 billion Austrian schillings—at present value about a billion dollars. It is rated at 700 megawatts, about two-thirds the size of Seabrook and Shoreham, two American nuclear plants that are also ready to go and hotly contested.

"When Zwentendorf began, we didn't know anything," an Austrian environmentalist told me. "Nuclear power sounded better to us than a coal plant or another hydropower dam on the Danube. If only we had known then what we know now."

They know now that two of the four German plants with the same design as Zwentendorf have been shut down permanently by mechanical problems. They know now that Zwentendorf is located squarely on an earthquake fault zone. And during a Danube flood, water seeped into its containment vessel, so now they know that the groundwater is not protected from contamination in case of a meltdown.

Furthermore Austria, like every other country with nuclear power, has no plan for the disposal of nuclear waste. The original idea had been to bury it in deep granite under the Alps. But the villages at the chosen site vehemently rejected this plan, and by Austrian law a locality cannot be forced to accept such an imposition from the federal government. The Austrians offered the waste to Hungary, Egypt, and China, but all refused. The Shah of Iran was eager to have it, but then he fell from power. The Ayatollah wasn't interested.

By the time Zwentendorf was finished, so many doubts had been raised that the government was forced to hold a referendum to decide whether to start the plant. During the weeks preceding the vote, the argument raged—the same one that polarizes every country that permits public discussion of nuclear power. People were told they had to choose between progress and safety, between jobs and the environment, between present brownouts and future contamination. Bruno Kreisky, then chancellor, declared Zwentendorf a top priority and appealed for a yes vote. Austrians still do not agree whether he caused more anti-Zwentendorf pro-Kreisky people to vote yes than he did pro-Zwentendorf anti-Kreisky people to vote no.

At any rate, on November 5, 1978, 50.5 percent of the voters said no to Zwentendorf. The Austrian nuclear power program came to a halt.

In spite of the warnings from the pro-Zwentendorf side, there were no brownouts, and the country's economy did not collapse. The plant sat there, the focus of an impasse between the government, which never got around to decommissioning it, and the environmentalists, who suspected the authorities of waiting for a favorable political climate to start it up. Opinion polls on Zwentendorf continued to show about 50 percent for and against. There were rumors that the plant had been secretly tested and thereby contaminated with radioactivity.

Then came Chernobyl. Austria's first spring lettuce and strawberries had to be dumped, and farmers were only partially compensated. Children were not permitted to play in outdoor sandboxes. Austrians watched all the governments of Europe give conflicting, confusing statements about nuclear safety. The polls suddenly showed only 10 percent of the voters in favor of opening Zwentendorf.

The government says that now Zwentendorf will be decommissioned. No one knows what will be done with it. It could be converted into some other kind of power plant, but that would be nearly as expensive as building a new plant from scratch. Some have suggested filling it with the hazardous chemical wastes of Europe. Others want to use it as a conference center for antinuclear groups.

It used to be a joke that Zwentendorf could be a museum for failed technologies, with a wonderful one-to-one scale model of a nuclear power plant. The province of Lower Austria has taken the joke seriously. It would like to make a park there, called "Historyland," an exhibition on the development of human civilization.

Is Austria the only country that can walk away from a completed nuclear power plant? Could a Zwentendorf-like story apply to Seabrook and Shoreham and the other twenty-six nuclear plants about to come on-line in the United States? Here the financial consequences of a cancellation fall on private investors rather than the public budget. Here the residents of whatever site the government chooses for waste disposal have no legal recourse. We have, as Austria does not, a domestic nuclear industry and a nuclear-based defense system with powerful lobbies. And we do not decide the fate of nuclear power plants by public referenda.

If we did, polls indicate that 78 percent of us would vote no. The utility companies themselves are voting no. The last twenty-eight plants in the pipeline were all ordered before 1974; all subsequent orders have been canceled.

The Austrians are not the only ones who have learned that whether or not it is an obsolete technology, nuclear power is an expensive and dangerous one. The question is whether they are the only ones with a political system flexible and democratic enough to respond to what most of the people have finally learned.

Why Nuclear Power Is Not the Answer to the Greenhouse Problem

To THE folks who bring us nuclear power, the hot summer of 1988 was a bonanza. They didn't have to advertise any more; the media were doing it for them. Nearly every report on the drought called it a sign of the global greenhouse effect. One cause of greenhouse warming is carbon dioxide from burning fossil fuels. Nuclear power produces no carbon dioxide. Therefore, the logic goes, to prevent more droughts, we should nuclearize in a big way.

The problem with this logic is that it doesn't hold up when you run out the numbers. Bill Keepin and Gregory Kats, energy analysts at Rocky Mountain Institute in Old Snowmass, Colorado, have figured out what it would take for nuclear power to replace fossil fuels. Their numbers suggest that there is a solution to the greenhouse problem— but it isn't nuclear.

Suppose the world's nations agree to replace all present and future uses for coal with nuclear power and to do so within forty years. (Coal, rather than oil or gas, because coal is the greatest carbon emitter and because nuclear can substitute directly for coal in making electricity.) Keepin and Kats make the deliberately optimistic assumption that each nuclear plant can be built in only six years and at a cost of $1,000 per kilowatt capacity. (The figures are for France; in the United States nuclear plants take twelve years and cost $3,200 per kilowatt.)

If world energy demand grows by 3.5 times between now and 2025, Keepin and Kats calculate that substituting nuclear for coal would require 8,000 large nuclear plants, as opposed to the 350 operating worldwide today. New ones would have to come on line at an average rate of one every 1.6 days, at a cost of $787 billion per year, for thirty-eight years.

Even with this enormous shift to nuclear power, carbon dioxide emissions would grow to be 65 percent higher than they are now.

Greenhouse warming would be rampant. The drought of 1988 would look like a cool spell.

If energy demand goes up more slowly, doubling by 2025, and again nuclear is systematically substituted for coal, one new nuclear plant would be required every 2.4 days, at a cost of $525 billion annually. There would be eighteen times as many nuclear plants as there are today. Carbon dioxide emissions would grow until the turn of the century and then slowly fall, but they would always be higher than they are now. The greenhouse effect would go on getting worse.

Nuclear power is ineffective in combatting greenhouse warming because it provides only electricity, which accounts for just one-third of fossil fuel use. Fossil fuel use accounts for only about half of the greenhouse problem (the rest comes from deforestation and from gases other than carbon dioxide). And nuclear starts from too low a base and takes too much time and money to replace coal quickly.

Even if the managerial capacity were available to construct so many plants so fast, the drain of capital into nuclear construction would slow the very economic growth that is assumed to require so much power in the first place. And of course the problems of nuclear power—dangerous wastes, threats to public health, decommissioning, diversion of materials into bombs, vulnerability to terrorism—all would escalate.

Now for the good news. There are ways to ameliorate greenhouse warming. They involve state-of-the-art design to meet energy needs efficiently. That doesn't mean cold rooms, warm beer, and general deprivation. It means being smart about warming rooms and cooling beer so as to use the least possible amount of energy.

Unlike nuclear, efficiency improvements are fast and cheap, and they apply to every kind of energy use, including transportation. Changing all the lightbulbs in America to the most efficient ones now available could shut down forty large coal-fired power plants and save the nation $10 billion a year. If office buildings were constructed in the most energy-efficient way, they would save the equivalent of eighty-five power plants and two Alaska oil pipelines, at no increased cost. If the fleet efficiency of U.S. cars doubled from the present 18 miles per gallon to 36, automobile carbon emissions could be cut in half (another half if the fleet reached the 78 mpg of some current five-passenger full-size test vehicles). As a side benefit there would be huge reductions in urban air pollution, acid rain, and Persian Gulf military costs.

According to one study, with a major commitment to energy efficiency the industrialized nations could maintain GNP growth rates of 1–2 percent per year and cut per capita energy use in half. The developing countries could grow even faster and keep per capita energy demand nearly constant. Carbon emissions would decline. Add a shift to solar energy and reforest the earth, and the greenhouse problem could be reduced dramatically.

Some folks are so bedazzled by nuclear power (usually because their own finances or status depend on it) that they can think of only one answer to any energy problem. They understate the limitations of nuclear, they assume it can produce miracles without stopping to calculate costs, and they don't see other choices at all.

Those of us with no stake in the nuclear industry can be more rational. We don't have to choose between living near a nuke or turning our grainbelt into a desert. We can see other solutions to the greenhouse problem and start with the quickest and cheapest ones first. At current costs a dollar spent on energy efficiency displaces nearly *seven times* as much carbon as a dollar spent on nuclear, does it sooner, and does not generate long-lived radioactive wastes. The same dollar can't be spent twice. It makes sense to spend it on efficiency.

Solar Hydrogen—a Fuel for Avoiding a Greenhouse Future

I F GEORGE Bush were an environmental leader, he would see that the greenhouse problem is in fact an opportunity—to move beyond fossil fuels, oil spills, air pollution, and enthrallment to the Middle East, toward clean, renewable energy that we can tap on our own territory.

If he were at the entrepreneurial edge not only of environmental policy but energy policy (hey, I can dream, can't I?), he would make this nation a leader in developing the fuels of the future. Chief among those fuels is hydrogen made with the energy of the sun.

The World Resources Institute (WRI) recently released a report by Joan M. Ogden and Robert H. Williams that summarizes the potential of solar hydrogen. It makes clear that solar energy research has been making quiet but impressive progress.

Photovoltaic cells generate electricity from sunlight. At present they do so at a cost of 10–11 cents per kilowatt-hour (the rate I pay to my electric company here in the expensive Northeast). The way its costs are falling, photovoltaic electricity should be available for 2–3 cents per kilowatt-hour by the year 2000, say Ogden and Williams.

That would make it competitive with other electricity sources, except for one small problem. The sun doesn't shine all the time. In some places, like the expensive Northeast in November, it hardly shines at all.

That's where hydrogen comes in as a way to store and transport solar energy. Photovoltaic electricity can be used to split water into hydrogen and oxygen. The oxygen goes off into the atmosphere. The hydrogen is shipped through pipelines like natural gas, or compressed into bottles to be burned when and where it's needed.

On burning it recombines with oxygen to form water again. It releases no greenhouse gas, no sulfur dioxide, no particulates, ozone, volatile organics, no pollutants at all except some nitrogen oxides (which are formed whenever there's burning in our nitrogen-rich atmosphere).

As opposed to massive coal or nuclear power plants, which take years to build and billions to finance, photovoltaic hydrogen rigs could be set up quickly in small units. It would make sense to put them mostly in the sunny Southwest. If they were to replace all the oil, gas, and coal now used in America, they would cover about one-half of 1 percent of the U.S. land area, or 7 percent of the desert area. That amounts to about half the state of New Mexico, though, of course, the solar collectors wouldn't all be in one place. They could go on roofs or abandoned mines or other already-disturbed spaces.

Solar hydrogen for residential and industrial energy use would be technically easy to arrange. The hard part is portable energy for vehicles. The hydrogen gas would have to be compressed into cylinders or, more likely, stored as metal (iron, titanium, or nickel) hydride.

The metal in the cylinders can be recharged with hydrogen over and over for the lifetime of the car. A metal hydride cylinder would take about the same space as a 12-gallon gas tank and would carry enough energy to go 200–300 miles between refuelings—but only if the car is much more efficient than our current gas guzzlers.

Carmakers already know how to make cars that are efficient and that can burn hydrogen. The best engine for the job would be a stratified-charge injection engine. The company farthest ahead in developing a hydrogen automobile is Daimler-Benz of West Germany. They estimate the final fuel cost at about $2 per gallon of gasoline-equivalent, which sounds expensive to us but not to Europeans. That price would look cheap to us too if we were charged the full cost of the pollution our cars generate.

Hydrogen fuel would work even better for trucks, buses, trains, and other large vehicles where the weight and volume of fuel are less problematic. Airplanes would be the most difficult; at the moment no one sees an economical way to power them with hydrogen.

When I was a chemistry student, I generated enough hydrogen fires in the lab to be very respectful of this gas. So I was most interested in the section of the WRI report that compares the safety of hydrogen to gasoline and natural gas. They turn out not to be very different; the same precautions would be needed that we take with the fuels that already pervade our world.

Solar hydrogen is not a proven technology. Neither are nuclear energy, coal synthetics, gasohol, or any of the other exotic fuels into which our government has poured millions (in the case of nuclear, billions) of dollars. But for some reason the government is nearly blind to the hydrogen alternative. The United States spent $3 million on hydrogen fuel research in 1989. Japan spent $20 million, West Germany $50 million.

If George Bush believed in a free market instead of protecting powerful industries (I'm still dreaming), he'd remove the enormous subsidies of nuclear and fossil energy. He'd include environmental and social costs in the prices of present energy systems. He wouldn't have to do anything more. With a level playing field, there would be hot competition to develop a fuel that would generate no significant air pollutants and no hazardous wastes, that would be based on simple water, that would be tapped with collectors made from abundant sand, and that would never run out as long as the sun shines.

LAND USE
AND URBAN
GROWTH

EVER SINCE doing *The Limits to Growth* study, I have been fascinated with the power of the idea of growth in our culture. Growth is the motive force of our vision, it is the receptacle of our hopes, it is close to a religion with us. We do not permit anyone to question the goodness of growth, and yet we don't really know what we mean by it. Around the concept of growth we have assembled a fantastic array of myths, myths that could be easily disproved by a simple look around. Growth does not end poverty. Growth does not generate enough money to clean up the pollution caused by growth. Growth does not lower taxes.

It's especially poignant to watch the growth fable work in rural America, where I live. The officials of my typical small town talk constantly as though all our problems will be solved by growth. Yet the bigger towns around us, which have grown just as we say we want to, do not have lower taxes, better schools, more honest governments, less unemployment, or less crime. If anything, they have more problems than we do.

You'd think one could get up at a town meeting and say that and then the message would be clear. But I repeatedly try that experiment, and after a moment of uncomfortable silence, the conversation goes back to enhancing growth. It is a perfect, frustrating example of a paradigm being impervious to contradictory evidence.

So every now and then I try to put the argument in writing, to encourage those who do understand (there are many, they just never seem to be town officials) and who are working to define growth, to discriminate one kind of growth from another and to control growth for the long-term good of their communities.

The Only Sure Result of Growth Is Growth

C ITY COUNCILS and planning boards seem to be guided by one sacred belief: It's good to grow. Why? Because development will broaden the tax base and keep down taxes.

Everyone seems to know that's true, but everything in my personal experience says the opposite. When I moved from a big city to a small town, my property taxes went down and the quality of schools and services went way up. Since then the town has grown 50 percent bigger and my property taxes have tripled. So I have developed a sacred belief of my own: Growth costs dearly for the people who are already here.

I have a habit of testing beliefs against the evidence whenever I can. A few years ago I plotted the equalized tax rate for every town in New Hampshire against each town's ten-year growth rate in population and total assessed value. If growth really brings down taxes, those graphs should have sloped downward. If, as I thought, growth raises taxes, the graphs should have sloped upward.

The graphs looked like the stars in the New Hampshire sky. Pure scatter. In some towns growth was followed by lower taxes, in other towns by higher taxes. Neither sacred belief survived that test. It seems that the kind of growth and the way the town spends money are more important to the tax rate than the simple fact of growth.

To show how that works, here is a sample calculation of the direct tax impact of a development proposed in my town last year. The plan was for 60 houses on 280 acres. Let's assume first that the houses would be assessed at $100,000 each and that together they put 60 new children into school. At the current tax rate, those houses would bring in $127,000 in additional town revenue every year. But if our per pupil and per household costs stay the same, the new development would cost the town $142,000 in school costs and $44,260 in garbage pickup, road maintenance, and so forth. Those 60 houses would raise the taxes of everyone else in town a total of over $60,000 per year. (That comes to $30 for every man, woman, and child.)

On the other hand, if the houses were worth $150,000 each and put only thirty new children into school, taxes for the rest of us would go down by $75,000 per year.

These calculations do not take into account secondary effects—the necessity to build a new water system or school or the possibility of new stores or businesses. If a 60-job industry came to town instead of a 60-home development, roughly 60 new families would move in nearby (since we already have nearly full employment), but maybe they'd settle and put their kids in school in another town.

So, growth can lower or raise taxes, or growth in one town can raise or lower taxes in another town.

I don't know why we always talk about the tax implications of growth—which are close to unpredictable—when there are other implications far more certain. We can surely expect from that 60-house development 60 more families, at least 60 more cars on the road, 60 septic tanks, 60 water and electricity hookups, 60 chimneys emitting air pollution, 280 acres less of field and forest, a town that feels and acts a lot less "small town," some construction jobs for a while, and (the real motivating force) profit for the developer. The one certain result of growth is not lower taxes but growth.

Every town has a number of instruments with which it can control growth, to maximize its benefits and minimize its costs. They include zoning, subdivision regulation, land trusts, and conservation easements. We would use those instruments more effectively if we dropped the sacred belief that growth is always bad or good. We need to get much more specific. Growth of exactly what, exactly where? What concrete results will it have in the near term and the long term in our town and the town next door? Are those results fair? How can we distribute the costs and benefits of growth as fairly as possible?

Cherish and Protect Beautiful Places

MY WORK requires me to travel a lot. Wherever I have been in the world, no matter how nice it was, I feel, as the plane lowers over Grantham Mountain or the bus starts down the Enfield hill, overwhelming gratitude.

I am back home, in a place where Canadian highs bring in air so clean you take deep breaths just for the pleasure of breathing. A place where the streams and wells still carry mostly crystal-clear water. A place people come long distances to visit, just because it is so beautiful.

I appreciated this valley most of all during the first few months I lived here. I had spent the previous eight years in the scruffy edges of Boston where students can afford to live. For miles around me was a wasteland of industrial ugliness. I remember only one beautiful thing, a valiant wild cherry tree I passed on my way to work, surviving somehow, surrounded by asphalt, near a railroad crossing. That little tree was in bloom in May when I moved up north.

My soul had been conditioned to endure twenty blocks of red brick, neon, and shattered glass, waiting to drink in the sight of one cherry tree. Then I moved to an old farm in a green valley with thousands of blooming trees. Everywhere I looked there was beauty. I didn't have to wipe black grit off the windowsills every day. The water didn't taste like chemicals. The houses fit snugly on the hills, the people moved slowly enough to talk to.

I no longer had to hold myself in a state of internal tension, screening out ugly smells, sounds, and sights. I hadn't even been aware of that tension until I started letting it go, unfolding my senses again. I went around marveling at how full life can be, how alive one can feel, in a place that is simply beautiful.

But here, too, we are "developing." We have tangled arguments about this "development." Should we welcome it, regulate it, forbid it? Who has a right to do what? Is the Purity Supreme an improvement over the hayfield it replaced? If it is convenient to shop there, but its traffic adds an hour a week to my commuting time, plus a daily streak of ugliness, have I come out ahead? Does ugliness matter if there is a profit to be made?

These are difficult questions and we shouldn't give them knee-jerk answers, but we usually do. Some of us are just plain against development, all development, and some are for it, come what may. When we try to convince each other, we talk about jobs, profits, and taxes.

Jobs, profits, and taxes are important, but they are not really at the heart of the matter. The jobs are mostly for people we don't know, who don't live here, but who are expected to show up. Growth has yet to lower taxes noticeably in any place I know about. Profits in the new supermall replace the profits of the farms and the small stores the mall displaced; the profits tend to flow toward Connecticut or Boston now instead of staying in the valley.

I think we talk about jobs, profits, and taxes so much because they are countable things, *quantities*, and therefore legitimate considerations. Other considerations, such as cherry trees, grit on the windowsills, noise, neighbors, topsoil, and how it feels to be stuck in traffic by the supermall, seem not quite legitimate to talk about because they are *qualities*. It's tough to stand up in town meeting and talk about health, beauty, security, community, self-reliance, aliveness. It just isn't done, though in fact qualities are very real, and they are ultimately what count.

It is widely believed that "emotional stuff" is so uniquely individual that it can form no basis for public policy about development. I think that's true when the choice is trivial—should there be a K Mart or a Dunkin' Donuts on the bank of this stream? But when the questions are serious and the answers taken seriously, a community can speak with one voice on issues of quality.

Once I helped out with a door-to-door survey in my town just before we passed our first zoning ordinance. An overwhelming majority favored no road upgrades ("would just make more traffic"), no shopping malls ("we can already buy everything we need"), no building at all on prime agricultural land, and no big developments that would drop hundreds of new families into the town at one time ("we wouldn't know everyone any more"). But the town was fiercely in favor of young folks being able to put up their own houses or to start their own businesses. There was great sympathy with low-income people, who should, the town thought, be able to have a place, even a trailer, "but not keep it junky." Everyone could name a favorite, beautiful part of town that should be protected from all development—and they all named the same places.

The only dissenters were the half-dozen people in town who had plans to sell to big developers and dollar signs in their eyes.

People were surprised, touched, and excited to be asked these questions. Most of them had never dared to express their feelings in public because they were sure no one else thought that way.

You could see this response as rural selfishness—we've got ours and we won't let anyone else in—or you could see it as an expression of love for a place, a plea for appropriateness, and a recognition that the kind of development that respects the land and the community's needs is more likely to come from within than from outsiders who arrive, build, and then leave the town to cope with the results.

To make development decisions on behalf of the community, the land, and the future, we have to use a higher human faculty than the simple ability to count. We have to ask quality questions. Will this project, which will affect our valley for decades to come, demean the lives of the people it touches? Is it motivated by someone whose senses are shut down, whose purpose is to substitute material stuff for the emptiness of a shut-down life? If so, it will have an unrelenting ugliness that will deaden all who are exposed to it. If not, if there is some care and commitment, some sensitivity to the land and community, a development project can uplift the builders, the owners, those who live or work or shop there, and everyone who passes by.

People can tell the difference, if you ask them.

Land Protection Is More Than Champagne and Quiche

S OME PEOPLE here in New England say that land protection is a "champagne and quiche" issue. I disagree. I think land protection is an issue for hunters, fishermen, hikers, snowmobilers, or anyone who is a tourist or has a tourism-related job; anyone who has a groundwater well; anyone who uses paper or lumber, drinks milk or eats apples; anyone who pays taxes; anyone who likes the look of a forest or field better than the look of strip "development."

That doesn't leave out many of us.

In the twenty-minute drive from my home to work, I pass mostly forests and hayfields. I would pay at least $10 per year just for the pleasure of looking at them. I bet a thousand people who use that road feel the same way. The Flatlanders who come up from the city just to drive around pay much more than that for each trip.

As "development" runs rampant, the road is not only losing its beauty, it is also gaining curb-cuts, traffic, stoplights, and cops. We will soon need road improvements, not to mention a new school and a new dump, all paid out of taxes. It might actually be worth $100 a year to me as a straight economic proposition to keep those forests and fields from being "developed."

The farmers harvest from the land corn or hay or wood worth maybe $50 to $500 an acre a year. If they're good managers, that income will go on forever. The rest of us use the lumber and eat the cheese.

To the surrounding communities the forests are worth a bit in taxes, but their greatest value doesn't show up on a ledger. The forests cushion and absorb the rainfall, charging the aquifers from which we drink, storing and filtering runoff to keep the streams clear and their flow regulated. Forests are worth billions of dollars in water supply and flood insurance, especially to the people in the cities downstream.

To a "developer," the land is worth at least $1,000 an acre, cash down, much more for an acre near an intersection or by a lake or with a view. It is worth that only once.

From then on the economics of the land are changed. There may be utility for a homeowner, income from a shopping mall, revenues for a town. The money flow is greater, but that value no longer comes from the land. It comes from high-cost inputs—construction and maintenance, energy, labor, sewers, trash collection—all of which draw resources from land somewhere else.

If we let the market guide "development," we lose sight of most of the value of the land. The market sees only the one-time big profit of the "developer." It discounts the modest perpetual income of the farmer. It ignores the beauty for each bypasser. The market does not value the groundwater, does not foresee the flood, and does not take into account future taxes for sewers and schools.

To include and protect all kinds of land value, something has to be added to the market, something that expresses the long-term public interest.

On the community level zoning is one option. It can protect such vital functions as groundwater recharge and floodwater control. It helps keep one person's "development" from putting a tax burden on others.

But zoning can't pick out particular pieces of land that urgently need protection. It can be the kiss of death to farmland, which is often physically well suited for "development" and zoned that way. And zoning is subject to so many exceptions, revisions, political deals, and conflicts of interest that it requires constant vigilance from the citizenry to be effective.

Another tool for protecting land is the acquisition of development rights. New Hampshire, for example, has a small fund to buy development rights to farmland. The farmer who sells those rights continues to own the land, manages it, lives on it, earns money from farming and logging it, and can sell it. But the deed is restricted so that no owner, now or in the future, *ever* can "develop" it. The state acquires only one right in the deal—the right to enforce that restriction.

Land trusts, conservation organizations, and towns can also acquire development rights to land. They can do a lot with a little money because a purchase of development rights is less expensive than an outright purchase of land. The land remains in private hands, on the tax rolls, and marketable. And a whole new market in protected land is created. In that market prices are set not by "developers" but by farmers, at levels that reflect the yield of lumber or hay, which means at levels that farmers can afford.

About a year ago I was trying to find $150,000 for a local land trust to buy the development rights to 320 acres of New Hampshire land. I ran into a friend who was working for a land trust on Cape Cod where "development" is far advanced. Her trust wanted to protect a 200-acre parcel. To do it, she had to come up with $3 million.

That's what happens to the economics of land protection when it's done too late.

Controlling Growth by Controlling Attractiveness

IN WOODSTOCK, Vermont, everyone's mad about a highway. In other places the issue is a sewer system or a school. The real issue, of course, is growth. According to Jay Forrester's Attractiveness Principle (Forrester is a professor of systems analysis at MIT), there's only one way to control growth—control attractiveness.

It's hard to imagine a more attractive place than Woodstock, a well-off, picture-perfect, white-church New England town, a destination of tour buses in October, the leaf-peeper season. It has clean air, clean water, rolling hills, trendy shopping centers, good schools, and good access via two interstate highways to the rest of the world.

One of Woodstock's problems, however, is that its connection to those interstates is Route 4. Route 4 is a twisting, two-lane rural road; it's a nuisance and a hazard. During every economic upturn more condominiums and shopping centers appear around Woodstock. And Route 4 is increasingly overloaded. The 30-mile stretch east of Woodstock averages two accidents a week. Often traffic is too thick to go the posted speed.

So the highway engineers propose to expand Route 4 to four lanes for 31 miles.

That idea is unpopular, to put it mildly. Public hearings have attracted hundreds of people, all seething. They say expanding Route 4 will just create more growth, more traffic, more highways. They don't want to live in a four-lane type of town. Leave Route 4 alone.

Their analysis is absolutely correct. Route 4's traffic count is reaching threshholds that are triggering the interest of a whole new bunch of commercial developers. And according to the Attractiveness Principle, that's only the beginning of the story.

In a free society if any place is unusually attractive, folks will—no surprise—be attracted there. The most mobile people (the young, the rich, the well informed) will get there first. The place will grow until its attractiveness has been reduced by crowded highways, or unemployment, or scarce housing, or pollution, or just plain visual blight. (The most mobile people will have moved on by then.) When the place is no more attractive than anywhere else, then and only then will it stop growing. What else can stop it?

Woodstock can't choose to be more attractive than other places. It can only choose the ways it prefers to be unattractive.

The attractiveness of a place is a complex combination of climate, economy, amenities, and scenery. No one can define attractiveness exactly, but people make up their minds about it every day by deciding to move from Hartford or Boston or Westchester County to Vermont (that's the direction they're moving at the moment). Millions of human judgments weigh Vermont's clean air against Boston's job market and Manhattan's cost of living. The very different mixes of attractiveness and unattractiveness in those places may seem incommensurable, but people make their comparisons and move around until attractiveness evens out everywhere.

Honolulu has a perfect climate and a beautiful setting, but it's thousands of miles from anywhere, and ordinary people can't afford its housing prices. Boston has jobs, museums, concerts, horrendous traffic, and a polluted harbor. Woodstock has Vermont charm, clean air, cold winters, black flies, and Route 4.

The normal instinct of public officials, including those of Woodstock, is to fix problems and make their community perfect. The more perfect they make it, the more new people show up. What Woodstock needs to do, Forrester would say, is decide what kinds of imperfection it's willing to live with.

A crowded, unsafe highway? If that's unacceptable, then choose something else. Superrestrictive zoning, perhaps, or an absolute limit on new curb cuts, or higher property taxes (I know, they're already too high, but not high enough to stop people from moving in). Bad schools. Bad air. No jobs. Developments so ugly you might as well live in New Jersey. Some sort of whopping surcharge on those developers. Either Woodstock chooses its form of unattractiveness or the growth process chooses.

It takes a while to absorb the implications of the Attractiveness Principle because it turns conventional thinking upside down (Forrester is good at doing that). Its implications are not good news for the sort of people who live in Woodstock. The principle says you can't live in a privileged bubble of attractiveness unless you are perpetually young, rich, well informed, and on the move at the head of the attractiveness wave. It says that growth is your problem wherever it occurs. It says the only way to be sure of living in an attractive place is to be committed to the attractiveness of every place.

A Neighborhood Plans Its Own Development

A FEW DAYS after Christmas, neighbors from eleven rural households in Lyme, New Hampshire, gathered at Hank and Freda Swan's house for an easement-signing party.

Family by family, they put their names to twenty different documents, amendments to their deeds, limiting forever the number of lots that can be subdivided and houses that can be constructed on their land. Together the easements will protect about 400 acres of farmland, forest, stream banks, and scenic views along two miles of road.

Conservation easements separate the rights of land ownership into two categories, use rights and development rights. The use rights—to live on the land, farm it, log it, sell it—remain with the owners. The development rights—to subdivide, build, pave—are eliminated.

To be enforceable, a conservation easement has to be conveyed to another party, usually a land trust, a town, or a conservation organization. The Lyme easements went to the Society for the Protection of New Hampshire Forests (SPNHF). The SPNHF has the right to monitor the land to be sure the easement's restrictions are observed. It can take legal action if the restrictions are violated.

Easements do interesting things to land values. If you put a conservation easement on your land, you will likely reduce its market value. But you will probably increase the value of your neighbor's land.

So the neighbors in Lyme essentially gave one another easements for Christmas. The net result, however, was a loss in land value. Most simply took their loss as a charitable gift to the nonprofit SPNHF. Others couldn't afford to do that. Their easements were purchased by contributions from townspeople in Lyme.

But economics were a secondary consideration, the folks at the party told me. "I don't think any of us thought about money when we agreed to the easements. We did it for the land. I did it for my grandchildren, so this road will be as beautiful for them as it is now. After we knew we would do it, then we looked at the financial implications."

It's fairly common for a single landowner to adopt a conservation easement, but it's rare for a whole neighborhood to do it. When I asked how the Lyme project began, I got three different answers.

Some people said it started when Shine King put his farm up for sale. Developers began to calculate how many houses could be placed on the fields. The neighbors began to wonder how that land might be kept in farming.

Freda Swan was another starting point. She had learned about easements in the League of Women Voters many years before, and she had been thinking about protecting her property and asking some of her neighbors to join her.

But Freda says that Don and Keita Metz really got the ball rolling by putting an easement on their Connecticut River frontage.

However it started, there was a lot of activity between Labor Day and Christmas. There were neighborhood meetings. Vicki Smith, a land-protection specialist for SPNHF, sat down with each owner to design an easement that fit that person's desires and that land's capabilities. Freda Swan assembled land surveys, raised money for the purchased easements, and constantly injected energy into the process. An appraiser started figuring out the value of each parcel with and without development rights.

By Christmas nearly every easement was ready. The Shine King farm was divided into six large parcels, all protected by easements, some purchased by neighbors, some by the Lyme Hill and Valley Association, a local land trust. Shine is going to go on farming it. On the entire road just eight more houses will ever be permitted and four more apartments in existing structures.

At the party I asked the neighbors what they thought other people ought to know about their project. Don Metz said, "They should understand that we aren't all one kind of person. Some of us own 65 acres, some own 5. Some are retired, some are young couples just starting their families. Some are natives, some newcomers. This isn't just a project for old folks or yuppies."

Freda Swan added, "I wouldn't want to say that everyone should do this. Neighborhood easement projects should be in the interest of the whole community. This road was selected by the town as a scenic road. People hunt and hike here. There are historic houses, wetlands, farmlands. If you look at Lyme, where the main roads are, where the growth is occurring, you see that this is one area that should be protected. Not just for us, but for the whole town."

Another neighbor said, "People should know that it's hard work. You have to know the land, look at its potential, ask what's best for it and what will be best 100 years from now. It takes a lot of thinking."

"Why should *you*," I asked, "be the ones to dictate what will or will not happen to this land? Why should you have the right to limit development for future landowners?"

"Well, people who develop do the same thing; they dictate the future. They destroy options for farming, forestry, wildlife, peace and quiet. All landowners impose on the future of the land in lots of ways. We've just chosen a way that preserves its beauty and productivity."

"Actually, we feel pretty good about it. We think people in Lyme 100 years from now will feel good about it too."

GARBAGE, TOXICS, SEWAGE, WASTE

As THE landfills close, America is finally admitting that there's no "away" to throw things to. Our response to the waste problem is still directed too much toward the wrong end of the pipe—trying to figure out how to deal with waste rather than how to cease generating it. But human beings are above all creative creatures, and ingenious solutions to waste problems are popping up all over—some (such as recycling) much more long-term-sustainable than others (such as incineration).

I write about waste issues often, not only because the topic is politically hot, but because there are so many exciting solutions that one can actually go see working somewhere. If there was ever a solvable problem, this is it.

Incinerator Blues I: Do You Know Where Your Trash Is Tonight?

ABOUT A month ago I got a mailing from an environmental group, warning about the very incinerator to which my garbage wends its humble weekly way. It's a brand-new, waste-to-energy plant in Claremont, New Hampshire, built and operated by Signal Environmental Systems.

"Watch this plant!" the flier said. "This is the second incinerator in the USA to be fitted with a dry scrubber and baghouse filter. Ash testing from March to June 1987 has confirmed that the better the industry gets at controlling air emissions, the more toxic the ash becomes. The plant is converting *four* tons of garbage into *one* ton of hazardous waste."

My garbage? Toxic ash? Hazardous waste?

A week later a headline in the local paper said, "Latest Results Are In, Incinerator Ash Shown Cleaner." Shortly thereafter another headline said, "Incinerator Ash Testing Questioned." Some experts were saying that the tests weren't done properly, that they aren't very informative anyway, and that the ash should not be allowed anywhere near our landfills. Other experts were replying, in polite expert fashion, that that is a lot of balderdash.

When experts disagree, I always assume it's time for ordinary folks to find out what's going on. Anyway, I consider it a basic rule of good citizenship to Know Where Your Garbage Goes, and I had never laid eyes on my incinerator. So I paid it a visit.

Warning: The following material is an explicit description of how a modern garbage incinerator looks, smells, and works. If you would rather not think about such things, consider that wherever you live,

167

there's likely to be one of these plants near you—there are seventeen planned for New Hampshire and over 1,000 for the nation over the next ten years. An incinerator will almost certainly try to find a home near you.

The $25 million Signal plant is in Claremont's industrial-park-in-the-countryside, surrounded by corn, forests, and houses with blooming gardens. It looks like any new industrial plant, a huge, ugly, sheet-steel block of a building with a tall rust-colored smokestack. Nothing visible comes out of the stack.

All day garbage trucks pull up the drive and onto the weighing platform. The computer inside has memorized each truck's weight; it registers the weight of the trash and totes up the tipping fee (which started at $40/ton and has risen over three years to $76/ton). About 200 tons of garbage arrive per day.

The trucks drive onto the tipping floor, a massive concrete-lined hall that can hold a thousand tons of trash. Air is drawn into the tipping floor and through the boilers to reduce odor, but on the floor your nose tells you plainly that garbage is the business here. The trucks dump it, a bulldozer shoves it into impressive mountains, a front-end loader drops mouthfuls of it into the chutes that feed the two boilers. There is no separating, milling, crushing. Whatever we throw away—bottles, cans, Pampers—gets tossed, all mixed up, into the boilers. That's why this is called a mass-burn facility.

The boilers are like woodstove fireboxes, only much bigger. They have small, heavy windows through which one can see the enormous fires. Their floors are stair-stepped; the burning trash moves slowly downward and the ash falls through grates and is quenched with water. This "bottom ash" is seared, unburnable trash, mostly metal and glass. From its daily 200 tons of garbage, the plant produces about 60 tons of ash. The incinerator doesn't eliminate the need for a landfill; it just reduces the volume going there.

The heat of the boilers makes steam, which goes through a turbine to generate electricity. When fully fired up, the Claremont plant produces 4.2 megawatts, about 1/250 as much as a nuclear power plant.

The gases coming off the burning trash are full of nasty stuff, and that's why there are two antipollution devices on this plant, a scrubber and a baghouse.

The scrubber runs the gases through dry lime, which neutralizes the sulfur dioxide and hydrochloric acid that would otherwise cause acid rain. The baghouse is a series of asbestos-and-teflon bags, like huge vacuum cleaner bags. They catch about 75 percent of the fine smoke particles called fly ash. The fly ash contains most of the most toxic pollutants, including dioxins and heavy metals—lead, cadmium, mercury, chromium.

The fly ash from the baghouse is combined with the bottom ash and trucked to a plastic-lined ash landfill in Newport, New Hampshire, just 1,000 feet from the Sugar River, which is the drinking water supply for Claremont.

Is that ash safe? Will the pollutants in it blow around in the air or leach out into the groundwater? That's what the experts are arguing about. While they argue, the Claremont plant is burning away, creating 60 tons of perhaps-safe, perhaps-hazardous ash a day, under the all-American assumption that industries, like people, are innocent until proven guilty.

Incinerator Blues II: Getting the Lead Out of Our Ash

IF THE ash from the Claremont incinerator is deemed officially hazardous it must be placed in an expensive, lined, monitored, hazardous-waste facility (none of which exist in New Hampshire). If it isn't, it can go into any old landfill.

The difference will mean millions of dollars to the region, billions of dollars to the incinerator industry. It may determine whether that industry survives. As you might imagine, the agencies that rule on the hazardousness of incinerator waste are under a lot of pressure. Here's what they're fighting about.

The Environmental Protection Agency decides whether ash is hazardous by conducting an "extraction procedure toxicity test" (abbreviated as EP tox). In the EP tox test dilute acetic acid is poured over ash samples, then they are shaken and filtered. The filtrate is tested for pollutants, one of the most common of which is lead. If the lead concentration is more than 2 milligrams per liter, the ash is officially hazardous. Note that this test doesn't tell you how much lead is *in* the ash, only how much comes *out*.

The Claremont incinerator flunked this test for the first three months of its operation. The state of Washington tested ash from seven different incinerators across the United States and found all fly-ash samples hazardous under the EP tox test and all bottom-ash samples hazardous in tests for carcinogens (the ash from Claremont is mixed bottom and fly ash).

The EP tox test is supposed to simulate what would happen to the ash as water percolates through it for twenty-five years in a landfill. Of course the test *isn't* the same as what would happen in the landfill, and that is the core of the debate.

To help understand that debate, here are some more useful facts.

1. Heavy metals such as lead are indestructible. They are basic elements, and though they may undergo chemical transformations and may move about in the environment, they never disappear. They are never rendered harmless.

2. Heavy metals accumulate in the human body where they interfere with many functions. Lead causes neurological impairment, mental retardation, and hearing loss. Children are especially susceptible.

3. Whatever heavy metals are put into an incinerator come out, either in the stack gases or in the ash. The more you clean them out of the gases, the more they show up in the ash. The Claremont incinerator has lead in its ash because its baghouse is effective in removing it from the air.

4. Incinerators don't create lead, but they do make it more mobile than it was in the garbage. They concentrate it. They heat it up. Much of it vaporizes and recondenses in the scrubber and baghouse, adhering to fine particles of fly ash. These particles are easily blown around, inhaled, or washed away with water and into soil.

5. Heavy metals in the ash come from the garbage we throw away.

Primary sources of lead are batteries, solder, metal alloys, dyed plastics, and colored inks. Lead has been taken out of black newspaper inks, but it is still present in some colored inks. The only way to keep an incinerator from putting out lead is to stop putting lead into it, either by stopping the use of lead where substitutes are available (as in inks) or by recycling objects containing it (such as batteries).

A "lead-immobilizing" process used by Signal reduces lead in the EP tox test—not in the ash. The process (which is an industrial secret) probably affects the solubility of the lead by changing it to lead phosphate; therefore, the lead is less likely to wash out in the test. Whether it is less likely to wash out in a landfill is unknown. By the time we find out in twenty-five years, the Claremont plant will have generated and buried *500,000 tons* of ash.

The current discussion is about lead. In the future it might be about cadmium or mercury or dioxins or furans or other pollutants that come out of incinerators. The final fact worth remembering is that we are the ultimate source of these things. We might like to think the garbage disappears forever when the truck hauls it away from the curb, but this is, after all, a small planet, and materials don't leave it. They just cycle around. Every now and then they come back to haunt us.

Incinerator Blues III: Where to Put Your Trust?

A S I GO back and forth from the pro- to the anti-incinerator side of the controversy in my valley, I can't find any villains. What I find is likeable folks on both sides, who can't agree on where to put their garbage because they can't agree on where to put their trust.

Al Haley, the operations manager at the Claremont incinerator, is a wiry, energetic, can-do sort of guy. When he shows off his plant, he apologizes for the dust (I hadn't noticed any dust). The start-up period messed things up, he says, but he's got the problem licked, and the place gets cleaner every week.

Haley thinks the pollution-control procedures he has to follow are a nuisance, but he does them with care. He's an engineer; he leaves the environmental fight to others and does his best to comply with whatever regulations come along. He gets mad at people who say that no one who works for the plant will live near it; he wants it known that he lives right nearby. The plant's safe and it's performing well, he says with a shy smile, "because I'm running it."

Connie Leach, project manager for the Solid Waste District, is a long-time environmentalist. At Williams College she started a recycling program; she has a master's degree in natural resources from the University of Michigan. She says it's hard to be on the incinerator side of the fence, but that's the best place for a steward of the environment to be—inside, working on solutions, rather than outside pointing a finger.

Leach wants to supplement the incinerator with an active recycling program. She insisted that the road to the plant be paved with glassphalt from reclaimed glass. She has set up buckets for old batteries at stores, so people can recycle them when they buy new ones. Signal has negotiated "put-or-pay" contracts that require towns to pay for their quota of trash whether they send it to the incinerator or not, so there is a strong disincentive for recycling. But Leach says that as the towns grow, recycling will pick up.

Keith Forrester, environmental engineer for Signal, can't understand why everyone is so worried about lead in incinerator ash. Of course lead is dangerous, he says, that's why it's regulated. It's probably the most well-studied toxic substance in the world. The regulations have safety factors built in; the Signal plant meets the regulations. What's all the fuss about?

I can't resist poking to see if Forrester's trust in technologies, plant operators, and regulatory agencies is genuine. I mention uncertainties about lead testing, changes in the safety standards, mechanical slipups. He's unshakable. He responds with streams of technical explanations. Error, corruption, inattention, and plain old ignorance don't exist in his world.

Bill Gallagher of Working on Waste (WOW), the citizens' group opposed to the incinerator, doesn't trust regulators at all. "We never get anywhere if we go to a regulator," he tells me.

The Water Supply and Pollution Control people say the incinerator isn't *their* problem until groundwater is actually contaminated. Boy, does that make me feel protected! They're going to test quarterly, and the incinerator has to fail tests consistently before they'll take action. The plant could be polluting for a year before they'll admit there's a problem!

The air pollution agency is in one place, water in another, solid waste in another. There's no coordination, and none of them welcomes citizen intervention. The director of the environmental agency used to be an incinerator company executive. Now he's regulating them! We're supposed to *trust* this process?

The members of WOW include two carpenters, a farmer, a nurse, two housewives, and a retired steel executive. Their conversation is about EP tox tests, fly ash, dioxins, and permits. They've learned a lot about this plant that has impacted their lives. They spend their evenings having meetings and going to hearings. Their spouses keep asking when this trash business will end so they can see their kitchen tables, which are always piled high with reports. "The pastures are going downhill, the cows are jumping out, we're in debt to the lawyers," Gallagher's wife tells me, but she's not complaining. She's in the fight herself.

"Do you want a clean incinerator or no incinerator?" I ask the members of WOW. They want no incinerator. They want a recycling system run by citizens. They want to be dependent on themselves, not on experts, regulators, and distant companies.

Haley trusts himself to run a good plant. Leach trusts the ongoing process, as long as environmentalists become a part of it. Forrester trusts the industry and the regulators. Gallagher trusts the common folks. Each of them is basically saying, "Trust me and people like me. And don't trust those other guys."

Most of us in the valley are not involved in the fight. We lean back and generate the garbage and watch with amusement as these characters confront one another, lay procedural traps for one another, and seek by force and counterforce some acceptable way to handle our trash. Whomever we end up trusting, we owe some appreciation to our fellow citizens who are willing to engage, to take a stand, to put their lives, credit ratings, cows, and fences on the line, to make sure the issue of public trust is raised, again and again, for us to decide.

Incinerator Blues IV: Why Not Recycling?

M OST EVERYONE involved in the Great National Garbage Problem would agree on the following propositions:
1. Landfills are better (and more expensive) than open dumps; they reduce smell, rats, and pollution of air and surface water. But landfills take up space and pollute groundwater.

2. Mass-burn incinerators are better (and much more expensive) than landfills; they take less land and they recover some of the energy from the trash in electricity. But incinerators generate air pollution and toxic ash, for which they still need landfills.

3. Recycling is better (and much less expensive) than incineration; it reclaims nearly all the energy and the materials in trash, it reduces pollution not only at the dump but at the point of manufacture, it postpones the depletion of mines and wells, and it creates local jobs.

In spite of these rankings, the nation is rushing to incinerators. There are more than seventy trash incinerators now operating in the cities and towns of the United States. At least that many more are under construction or in advanced stages of planning. The industry projects that $35 billion will be spent to construct municipal incinerators over the next ten years.

Why incinerators? Why not recycling?

One reason is the enticing profitability of the incinerator industry. Tax-free municipal bonds provide the money to build the plants. Taxpayers pay operating costs through tipping fees. Towns arrange for hauling, supply the landfills to receive the ash, and even sign put-or-pay contracts to guarantee a given amount of garbage. And since the federal 1978 Public Utility Regulatory Policies Act, utilities must buy the electricity generated at a fixed price.

Capital guaranteed, input guaranteed, market for output guaranteed. Building an incinerator must be one of the least risky, most subsidized business ventures in America.

Profit is one reason for the incineration boom, but another reason, perhaps the most important, is the set of myths that stand in the way of recycling. Listen to any public discussion of the garbage problem, read any of the hundreds of current articles, and you hear the myths.

They are the statements supported by no evidence and preceded by "of course," as in, "Of course people will never be willing to separate their garbage."

They are the assertions that everyone has heard but no one has personally checked out, as in, "There's just no market for recycled materials."

They are the demagogic, thought-stopping slogans that have no meaning, as in, "Let the market take care of the problem. Keep Big Brother out of our garbage!"

The best way to recognize a myth is to put yourself into a world in which it's not believed. Go to Japan or Europe, for instance, where all our recycling myths are being disproved. Or pick up a copy of *Resource Recycling*.

Resource Recycling is the journal of the rising American recycling movement. Paging through a copy is a cheery, myth-shattering experience. The ads catch your eye first:

Wanted. Used glass containers! Glass is more valuable than ever. (Owens-Illinois, Toledo, Ohio)

The CD3000 Can Densor. For recyclers determined to make money the *easy* way. (CP Manufacturers, National City, California)

A race against time. Landfills closing. We'll run your MRF [Material Recovery Facility] from start to finish. (New England CRInc, Billerica, Massachusetts)

We're proving how beguilingly attractive recycled papers have become. Paper this good-looking at prices this attractive will leave you
spellbound! (Prairie Paper, Lincoln, Nebraska)

Then there are the articles on tire recycling, plastic recycling, composting, and successful municipal recycling programs.

For instance, 80 percent of the residents of Wellesley, Massachusetts,
voluntarily separate their trash and drive it to the Municipal Recycling
and Disposal Facility. The facility is at the site of a closed-down
incinerator, now a park. There are lawns and picnic tables. Local
politicians seek votes there; Girl Scouts sell cookies. There are containers for aluminum cans, glass, metals, newspapers, and a popular
corner for reusable items, where books, games, toys, appliances, and
furniture can be passed on to neighbors or picked up by Goodwill
Industries. Yard wastes are composted, firewood is cut up and given to
residents.

The facility employs a staff of seven, and it *makes* $20 per ton on
material sales, while *avoiding* $40 per ton in hauling and tipping fees for
the landfill 15 miles away.

Any issue of *Resource Recycling* makes me think that the main thing
in the way of large-scale recycling in this country is a collective
failure of imagination. We need to start a new set of myths, which
stimulate our creativity and which might be closer to truths. For
instance:

Give us a clear choice between being taxed $40 a ton or earning $20 a
ton, between having an incinerator in our neighborhood or a recycling
station, and see how willing we are to separate our garbage.

Charge a fair disposal fee at the point of manufacture of unnecessary
packaging, unreturnable bottles, unrecyclable plastics, and hazardous
chemicals, and watch how the volume of garbage goes down.

Guarantee American businesses a cheap, dependable source of recycled materials, and notice how many ingenious uses will be found for
them.

Get Big Brother out of the incinerator industry or make equal
subsidies available to recyclers and watch a truly competitive market
discover the lower cost, the greater safety, the savings in materials
and energy, the decreased pollution, the ultimate wisdom of recycling.

The New World of Plastics—Not
New Enough

THE WORLD of plastics is in a mess these days because it has made a mess. Polyethylene, polystyrene, polyvinyl chloride, and all the other polys are piling up on roadsides, in the ocean, and in landfills. They are likely to last several hundred years there after serving us for a few weeks or hours—if indeed they could be said to serve us at all. The primary role of many plastics is to catch our attention in a store, for which purpose they are garishly shaped and colored with, among other things, toxic heavy metals. Environmentalists say the dumps are filling up not with packaging but with marketing.

Until we noticed the dumps filling up, most of us never thought about the stream of plastics flowing through our lives—18 million tons each year, of which 6.5 million tons is packaging, and over 3 billion dollars' worth is plastic bags in which to throw the other plastics out. Now everyone has panicked. In February 1989 the American Paper Institute counted the following bills pending in state legislatures (not counting those at the federal level)—66 proposed bans on nonbiodegradable packaging, 12 packaging taxes, 74 source separation and recycling mandates, and 19 requirements that state governments purchase recycled materials.

It's not fair to blame plastics for our trash problem, says the industry. They make up only 4 to 7 percent of municipal solid waste (by weight—by volume it's more like 20 to 30%). But plastics are the focus of most legislation, perhaps because they are the fastest growing constituent of trash, because they are used for so many trivial purposes, and because they are so nearly immortal.

Immortality is one of the qualities that makes plastics useful, of course. They are impervious to bacteria, acid, salt, rust, breakage, almost any agent except heat, and some of them can even stand up to heat. If they didn't junk up our lives so, we would regard them as miracle substances—long, long hydrocarbon chains, crafted to take on any properties we want. Plastics can be transparent or opaque, hard as steel or pliant as silk, squeezable or rigid, moldable into any conceivable shape.

And, environmentalists would say, they are made from depleting oil and gas wrested from the ends of the earth, transported, spilled, refined in energy-consuming, hazardous-waste-generating processes, synthesized, and disposed of carelessly. They are messy from beginning to end. If we were properly charged the full human and environmental costs of our plastics, we would not eliminate them—they are far too useful for that—but we would treat those specialized molecules with the respect they deserve. We would not use them for a few days or hours and throw them out.

The standard environmental formula for dealing with precious but polluting materials is simple. Reduce, reuse, recycle, in that order, and then, as a last resort, dispose with care. Of course, the plastics industry makes money in the inverse order. It is looking for a way to keep us buying millions of tons of plastics each year—and to have them miraculously disappear when we throw them away.

Therefore, industry's favorite answers to the plastics problem are two: incineration and degradation.

As a descendant of petroleum, plastic burns beautifully. Like all hydrocarbons, it combusts into carbon dioxide, a greenhouse gas, plus a host of other pollutants. Some of them derive from additives such as heavy metals (which end up either in air emissions or incinerator ash). Others, like the toxic dioxins and furans, come from high-temperature reactions between hydrocarbons and chlorine. Polyvinyl chloride (PVC) releases so much hydrochloric acid when it burns that it corrodes incinerators. For that reason one incinerator manufacturer recommends keeping PVC out of incinerated trash.

Incineration can recapture a small fraction of the energy put into making plastics. Degradation doesn't even do that.

Degradable plastics come in two forms: biodegradable and photodegradable. The biodegradable kind mixes the long plastic molecules, which nothing in nature can digest, with starch, which microorganisms will happily munch away. Depending on the material strength required and the rate of degradation, the starch percentage varies, but it's usually something like 6 percent starch to 94 percent plastic.

On the side of the road a bottle or bag made of biodegradable plastic slowly falls apart into tiny shards of undegradable plastic. The bottle or bag disintegrates, the plastic is still there. Presumably it is inert and harmless, but no one really knows the implications of a world filled with plastic sand.

In a landfill, biodegradation happens slowly, if at all. Nothing degrades well in a landfill. William Rathje, an anthropologist from the University of Arizona, drills core samples from old landfills and finds intact food, paper, and cloth that are twenty years old. He can date the layers exactly because he can read the newspapers. Landfills are not compost heaps. They haven't the proper air circulation, moisture content, mixture of nutrients, or communities of microorganisms to encourage natural breakdown.

Of course, there's no sunlight in a landfill, either. Photodegradable plastics have chemical links built into their molecular chains that fall apart when hit by ultraviolet radiation in sunlight. The breakdown products are shorter chains, not so much plastic sand as plastic powder. If the plastic is polyethylene and the chains are short enough, soil organisms then appear to take over and digest them—the only evidence I've seen of real plastic biodegradation. But several months of photodegradation are necessary to begin the process.

Promoters of photo- and biodegradable plastics admit, when pressed, that neither can extend the lives of landfills one bit. They only help with the problem of litter. They are designer molecules to do away with the ugly evidence of our unwillingness to pick up after ourselves.

Recycling at least slows the waste stream and lets the plastics serve several times before discard. Only about 2 percent of the plastics we use are now recycled (as opposed to 29% of aluminum and 21% of paper), but that's not because it can't be done. Plastics are the easiest of all materials to recycle. Basically they just need to be shredded, re-melted, and reformed. The industry itself grinds and reuses 5 billion pounds of plastic scrap a year.

Two things are in the way of serious plastics recycling—separation and purification. Consumers can put paper and cans in separate waste containers, but they can't tell the difference between polypropylene and polystyrene. Bottlers have set up a voluntary coding system, which stamps resin types on 8-ounce and larger containers. But that does not help us separate the myriad other forms of plastics, including squeezable bottles with several kinds laminated together. (In a true recycling society such unseparable mixed-material containers would be banned.)

Some separation is easy, though. Some recycling centers now collect high-density polyethylene (HDPE) milk bottles and polyethylene ter-aphthalate (PET) soda bottles. They are not, unfortunately, made back into milk and soda bottles because of the problem of purification. One person in a thousand might have used a bottle to hold roach poison or kerosene before throwing it away, and contaminants could have per-meated the plastic—to come out later into the milk or soda.

Therefore, recycled plastics are not used to package edibles. Re-claimed HDPE and PET are made into carpet fibers, cushion stuffing, and scouring pads. Mixed plastics are made into lumberlike poles, posts, stakes, and slats for never-rotting barns, docks, fences, road markers, and pilings. These processes are better than nothing, but they are not real recycling. They will screech to a halt when we have as many immortal cushions and flowerpots as we need. The only recy-cling processes that can work in the long run are those that return materials to their same use—soda bottles to soda bottles, milk con-tainers to milk containers.

Reuse is preferable to recycling because it takes less energy and causes less pollution to wash out a coffee cup, say, and refill it, than it does to crush it, melt it, and reform it. Because of the contamination problem, most plastic containers cannot be reused commercially. They can be at home, however, because you know whether you've put roach poison in a cup or not. A real environmentalist would never use a plastic hot cup only once.

But then a real, real environmentalist wouldn't drink from a plastic cup at all. He or she would remember that the world once worked fairly smoothly with washable china cups. Recycling is better than disposal, reuse is better than recycling, but reduction is best of all. It's easier to deal with a flood by turning it off at its source than by inventing better mopping technologies.

Many European countries, which have had to confront the finite-ness of their landfills sooner than we have, simply ban PET soda bottles. Furthermore, they require glass bottles to be made in standard-shaped half-liter and liter sizes, so any bottle can be refilled by any beverage company. Europeans cheerfully return glass bottles in handy, reusable (plastic) cases and do not seem to regard it as an infringement of their basic freedoms.

Solutions like these—reduction solutions, solutions that distinguish the plastics we need from the plastics we don't need—will not come from industry. They will come through the political process and through the market, as we finally charge ourselves, one way or another, the real cost of producing and disposing of plastics.

When Is a "Cleanup" Not a Cleanup?

EXXON HAS JUST HIRED 100 more people at $16 an hour to "clean up" the massive oil spill near Valdez.

The government estimates that "cleaning up" the radioactive contamination at its weapons plants will cost over $100 billion.

The states have submitted a list of 2,000 hazardous waste sites for priority "cleanup" under Superfund.

What does "cleanup" mean? When the technicians scour the oil-coated shore or go behind skull-bedecked warning fences, put on moon-suits, and rev up their vacuum-cleaner trucks, what, exactly, is going on? What is left when they're done? Where do they empty the vacuum-cleaner bags?

I've been asking those questions for months. The answers I get depend on what site I'm asking about and what kind of hazardous substance. About the only firm conclusion I've come to is that "cleanup" is rarely the right expression for what we do with toxic messes. "Wall off" might be more accurate, or 'immobilize for a while," or "put somewhere else," or "hide."

For example, the number one site on the Superfund list, the Lipari Landfill in New Jersey, is an 8-acre dump containing 3 million gallons of mixed chemicals. Fumes from the dump brought tears to the eyes of neighbors; a lake downstream occasionally turned purple or orange. The "cleanup" plan called for "containment." The site was surrounded by an underground concrete wall and covered with a cap of clay and plastic. Chemicals continue to leach out in the groundwater, and fumes still come through the cap.

Now the site will be "recleaned." The lake will be dredged, the bottom mud put somewhere else, and the landfill will be flushed with water for seven years. The leachate from the flushing will be collected, treated, and put somewhere else.

Capping, containment, and putting somewhere else are the cheapest options, and the most common, as indicated by these Superfund case study summaries by the Office of Technology Assessment:

Unproven solidification technology was selected to treat . . . highly contaminated subsurface soil. . . . The cleanup will leave untreated contamination on the site. (Chemical Control Corporation, Elizabeth, NJ)

Capping was called a cost-effective, permanent cleanup even though it does not provide permanent protection. . . . Treatment of contaminated groundwater is not yet planned. (Compass Industries, Tulsa County, OK)

Pumping contaminated groundwater and capping the site were chosen instead of . . . excavation and treatment of contaminated soil. Water treatment cannot remove all the diverse contaminants at the site. (Conservation Chemical Company, Kansas City, MO)

A report by the Hazardous Waste Treatment Council estimates that only about 8 percent of the remedies selected for Superfund cleanup use the maximum technical option required by law. And maximum treatment at its best is still not "cleaning up."

In one of the best Superfund operations, PCB-contaminated electric equipment abandoned in a warehouse in North Carolina was hauled to a special facility in Alabama. The material was shredded, packaged, and sent to an incinerator in Chicago. In the flames the PCBs became water vapor, carbon dioxide (a greenhouse gas), and various chlorine compounds (causes of smog and stratospheric ozone depletion). The ash went back to Alabama to be landfilled. Much cleaner, but not clean. And very expensive.

For some hazardous materials "clean" is a possibility; for others it's not. Here's the best we can hope for:

- Strong acids and bases can be neutralized.
- Cyanides can be made harmless by chemical reaction.
- Organic chemicals, including petroleum and pesticides, can be digested by natural organisms in a few cases. Some fall apart when exposed to sunlight. All can be broken down into water, carbon dioxide (a greenhouse gas), sulfur dioxide (a cause of acid rain), and chlorine compounds (some toxic or carcinogenic), if carefully incinerated at the right temperatures.
- Radioactive elements cannot be "cleaned up." They must be strictly sequestered from wind, water, and all life forms until they break down naturally according to their internal atomic clocks. Their half-lives (the time it takes them to decay to one-half their original quantity) vary—12.5 years for tritium, 33 years for cesium-137, 24,000 years for plutonium-239, 4.5 billion years for uranium-238.
- Heavy metals such as lead, cadmium, and mercury last forever, no matter where you put them.

Even those "cleanup" efforts that are chemically possible are foiled if the toxic materials become dispersed in soil, carried off in water, absorbed into living things, wafted away in the air, or mixed up together. Scooping the spilled oil out of the water and off the rocks of Prince William Sound is, let's face it, simply impossible.

I suggest that we stop using the word "cleanup" when referring to hazardous materials and start facing the fact that we are making uncleanable messes. Then, maybe, we'll start handling dangerous chemicals with the seriousness they require.

A Hazardous Waste System That Works

I N LOW-LYING Denmark people do not live very far from their groundwater. Whatever they dump onto the land ends up in their wells, and quickly. Back in 1970, before Love Canal, before Times Beach, before any hazardous waste disaster in their own land, the Danes were working out how to prevent such disasters. By 1975 they had instituted a hazardous waste disposal system that is one of the best in the world.

The system is overseen by the central government, but its day-to-day operation is in the hands of cities and towns. Each municipality has a collection point to which households bring solvents, pesticides, used oils, anything they don't want to find in their water. Hazardous materials can also be taken back to where they were sold—unused medicine to the pharmacy, half-empty paint cans to the paint store, dead batteries to the hardware store. The stores take them, separated and labeled, to the collection point.

From the collection point they are trucked to one of twenty-one transfer stations, none more than 30 miles away.

Every industry must tell the municipality within which it operates exactly what types and quantities of wastes it produces. (That's a step we are just reaching now in the United States.) Unless the town or city gives a permit for on-site treatment, the industry must deliver its wastes—again separated and labeled—to the transfer station.

From the twenty-one transfer stations all materials go to a central facility called Kommunekemi. Kommunekemi maintains a professional staff that directs each type of waste to its proper treatment process.

About one-fourth of the materials arriving at Kommunekemi are relatively nontoxic and immobile. They are sent to a lined, monitored landfill.

Cyanides are destroyed by chemical reaction. Organic chemicals, solvents, and oils are burned in high-temperature incinerators, which provide steam to heat 45 percent of the houses in the nearby town of Nyborg. The smokestacks are fitted with pollution-control devices. Air emissions are carefully monitored. Ash from the incinerator is sent to a separate, labeled compartment of the same landfill.

Heavy metals are also sent to separate and labeled landfill sites, with the intention of reclaiming them some day. They are covered regularly with lime to maintain low acidity—so they will not leach into groundwater—and with a plastic membrane.

If you're noticing the frequent repetition of the words "separate and labeled" here, that's one of the keys to the Danish system. Noxious materials mixed together in a toxic brew are impossible to deal with. Separated materials can be treated as far as chemically possible.

The Danish Environmental Protection Agency keeps a permanent inspector at Kommunekemi to report any problems. There have been a few minor spills—broken drums and gas releases—but none has done measurable harm to the groundwater or to human health. For a facility that handles 100,000 tons per year of hazardous materials, the safety record is outstanding.

Who pays for Kommunekemi? Municipalities and companies are charged by the amount and type of waste they send. The fees are high, not only to pay for handling, but to encourage the reduction of waste. The chemists and engineers at Kommunekemi will consult with any industry or town about how to recycle materials or reduce wastes.

They consult outside Denmark, too. Says Per Riemann, Kommunekemi's manager, "So many people come here to see how the plant works that we have thought of charging a fee. Instead we have set up a worldwide consulting company to help other places design hazardous waste treatment facilities. Denmark has few natural resources, but knowledge is one thing we can export."

Danes are no more angels than the rest of us. Their system is neither perfect nor perfectly well accepted. They could recover and reuse more materials than they do. Despite the high disposal fee, the generation of toxic waste has been growing in Denmark by 17 percent per year. The people of Nyborg are resisting a needed expansion of Kommunekemi in classic Not-in-My-Backyard fashion.

Still, compare their situation with ours. We have at least 10,000 mixed, unlabeled hazardous waste dumps. The EPA estimates that 85 percent of our toxic wastes are still disposed of in an environmentally unsound way—12 percent directly into watercourses. We have a lot to learn from the principles of the Danish system, if not its details.

Its principles are simple. Put information and control at the local level, with the people most likely to be impacted by improper disposal. Make the system easy for everyone to comply with. Do not tolerate noncompliance. Catch wastes at the point of generation. Keep them separated. Hand them over to professionals. Charge a disposal fee high enough to ensure top-quality treatment. Encourage waste reduction.

Seems like something almost any country could do.

Ben & Jerry's & Solid Waste

BEN & JERRY'S is known all over New England not only for outrageous ice cream but for fun and funkiness. Started in a renovated gas station in Burlington, Vermont, by two ex-hippies from New York, the company has always done business—well—let's say differently.

All 360 employees come together once a month for idea-generating sessions. They've formed a Joy Committee that provides, among other things, backrubs for employees on stressful days. Ben and/or Jerry tend to show up and give away ice cream at community gatherings for good causes. The company puts 7.5 percent of its pretax profits in a foundation to further positive social change, with an emphasis on the environment, children and families, disadvantaged groups, and peace.

Even with the purest intentions, however, when you make over 4 million gallons of ice cream a year, you generate a lot of waste. When Gail Mayville was hired as Ben & Jerry's office manager in 1986, her first challenge was wastewater.

Ice cream equipment has to be washed scrupulously after each batch. That results in great quantities of milky, sugary, eggy, soapy water, more than the sewage treatment system could handle in Waterbury, Vermont, where the Ben & Jerry's plant was located. When Mayville arrived, she was told she had exactly one week to reduce the company's wastewater input to the Waterbury system.

"I started phoning," she says. "In five and a half days I had a solution." The solution was a local farmer who had just sold off his dairy herd. Mayville offered to buy him 300 pigs if he would feed them ice cream washwater. Some of the water would also be applied as soil nutrient on his 350 acres of fields. The farmer liked the idea. He had always been kind of partial to pigs.

With daily diet supplements of Heath Bar Crunch and Dastardly Mash, they must be the world's happiest pigs.

That was just the beginning of Mayville's waste reduction program. She is taking on one waste stream at a time. She recaptures office paper, turns it unused side up, and has it made into scratchpads. She bales up 9 tons of cardboard packing boxes a week and sends them to be recycled. She was bothered by the 500 large polyethylene buckets Ben & Jerry's discarded weekly until she found a company in St. Albans that takes in plastics, cleans them (with a maple-sap-bucket cleaner), and shreds them for recycling.

Mayville wanted to use unbleached, recycled paper in the office, but she discovered that its sources were too far away and special trucking too expensive. So she's working with a Vermont business consortium to organize enough purchasing volume either to bring the transportation price down or to convince a local paper maker to make recycled office paper.

Then there's the packaging.

Like many items on the grocery shelves, the Ben & Jerry's Brownie Bar twin-pack is almost as much package as product—two ice cream brownies, each wrapped in polyethylene, sitting in a plastic divider tray, the whole business inside a cardboard box. Ben & Jerry's is now experimenting with cellophane on the wrapping machines instead of polyethylene. (Cellophane is preferable because it's made from plants, not oil, and it's biodegradable.) The company is also trying out in place of the plastic tray a simple egg-carton-type cardboard separator, made, of course, from recycled cardboard.

Mayville's latest quest is to find a tamper-proof seal for ice cream pints that is not made from polyvinyl chloride (PVC). PVC is the most polluting of plastics, both in its manufacture and its disposal, but it's perfect for shrink wrapping. Cellophane won't work, Mayville has found, because it melts in the shrink wrapper's heat. She's trying out a parchmentlike paper band now.

By the time she read this far in the first draft of this column, Mayville was getting uncomfortable. "Don't make it sound like I have all the answers," she said. "I'm just getting started. And don't talk so much about me. I couldn't do anything without the cooperation of everyone at Ben & Jerry's."

Those are just the points I'd like to make. Environmental problems aren't solved by grand answers. They're solved by working at little problems, one by one, with determination and creativity. And they're solved when there's a lucky combination of a dedicated individual working in a willing, responsive organization, the kind of organization that not only gives part of its profit to peace but also recognizes, as Mayville says, that "we've got to have a planet to have peace on."

Different Approaches to Sewage Treatment—and to the World

JUST BELOW a bank of condominiums at the Sugarbush ski resort near Warren, Vermont, you can find three, sometimes four, sewage treatment systems being tested side by side. The story unfolding there is about more than the chemistry of sludge. It is about the mindsets and values with which human beings attack environmental problems. It's about, if you will, the Future Relationship of Human Beings, Technology, and Nature on This Planet.

Sugarbush has spawned condominiums, restaurants, sports centers, and other profitable sewage-producing entities built on bedrock. They are most heavily populated when temperatures are below zero, days are short, and biological processes work slowly, if at all. This place is the ultimate test site for sewage treatment schemes. Whatever works here will work anywhere.

And just about everything is being tested because Sugarbush is desperate. Rice Brook, a small trout stream that flows down the mountain, is contaminated in winter with ammonia from Sugarbush sewage. The state hit the resort with a $50,000 fine and a moratorium on any further sewage hookups. If the problem isn't fixed, the state could deny the resort a discharge permit, effectively shutting it down.

Sugarbush's current sewage system is a collection of settling lagoons, which flow into a flocculator that adds aluminum to precipitate out phosphate, followed by a chlorinator, followed by a leachfield (a bed of gravel through which the sewage drains), which ultimately percolates into Rice Brook. It is a sophisticated system of the traditional, land-intensive, out-of-sight/out-of-mind variety. By Vermont's strict standards, it doesn't handle suburban-density housing on a cold mountainside.

The most high-tech alternative Sugarbush has tried is reverse osmosis. In this procedure the sewage is shoved under pressure through a semipermeable membrane. Water goes through; everything else stays behind. "A terrific procedure—it blew everything away! I almost drank it," says a technician at the plant. Reverse osmosis is Space Age sewage treatment at Space Age prices. It generates a concentrated "brine" of stuff that doesn't go through the membrane. The company that sells you the system doesn't tell you what to do with the brine. Sugarbush has decided that reverse osmosis is too expensive and too briny.

There are two remaining contenders.

The first is a squat, square, windowless concrete structure with a sign at the entrance reading "DANGER CHLORINE GAS—turn fan switch on before entering." Hanging on the wall is a chlorine detector gauge, a gas mask. and a set of instructions "IN CASE OF A CHLORINE EMERGENCY." Inside is a maze of pipes and dials, gas cylinders, and reaction chambers. Bags of dry sodium hydroxide are piled up, each one stamped "DANGER CAUSTIC." This is a breakpoint chlorine plant.

Across the driveway is an arched, plastic greenhouse. Inside, under a network of walkways, is a greenish pool with air bubbling through it. The pool is sewage, but the place smells good, like a greenhouse, humid and fertile. Pots of geraniums are in bloom, and rafts of willow and eucalyptus float in the pool. At the far end is a lush marsh— bamboo and cattail, marsh marigold and swamp iris. There is only one warning sign, and it was put up for a joke—"NO DIVING." This is a solar-aquatic plant.

In the breakpoint chlorine process the effluent is made alkaline with sodium hydroxide and then blasted with chlorine gas. The chlorine oxidizes ammonia to nitrogen gas, which bubbles off into the atmosphere. Excess chlorine is inactivated with sulfur dioxide to produce sulfate and chloride. Then the whole business is filtered through activated carbon to remove any remaining chlorine.

In case you didn't follow all that chemistry, one of the operators summed it up, "We make some wicked, wicked water here!" Wicked because this plant must handle ammonia levels ten times higher than usual, so the chemicals are at very high concentrations. Wicked because the input chemicals are dangerous, and one possible byproduct, chloramine, is a carcinogen. But wicked in its intermediate steps only. If everything works right, at the end of the process 95–99 percent of the ammonia is removed, and the effluent contains nothing worse than salt and sodium sulfate.

The breakpoint chlorine plant comes from a pipe-and-valve mentality: "What chemicals can we use to get rid of ammonia?" The solar-aquatic plant comes from an ecological mentality: "How does nature handle ammonia?" It sees sewage not as a waste to be gotten rid of but as a resource to be cycled back into life.

Nature handles ammonia by turning it into fertilizer. Normal soils and waters are full of bacteria that transform ammonia into nitrate. Nitrate is taken up into plants, the plants are eaten by animals, the animals excrete ammonia again. That's the nitrogen cycle, one of the great natural flows of the planet.

Over at the solar-aquatic plant John Todd of Ocean Arks International of Woods Hole, Massachusetts, the company that's doing the research for the system, is punching holes into a styrofoam sheet. He sticks willow cuttings into the holes and then floats the sheet on the river of sewage. "I've heard that the Chinese do this," he says. "The willows put down roots and draw the nutrients out. The leaves make animal feed. I'm trying three kinds of willow to see which works best."

As raw sewage enters the greenhouse, it flows first through a cylinder of nitrifying bacteria gathered from Vermont ponds, then into the raceways where algae multiply in the water, taking up nutrients. Freshwater shrimp eat the algae. Bass and trout in aquaculture tanks at the purified end of the system eat the shrimp.

Todd lifts a corner of a styrofoam float. It's covered with snails and transparent globules of snail eggs. "Here are the hard workers of this place. They clean up the sludge. We drained the plant and found almost no sludge on the bottom." The snails are also fed to the fish.

The river of effluent takes five days to wind from one end of the greenhouse to the other. When it reaches the end, it is filtered by the marsh. The plants there have commercial value (watercress), or pretty blooms (marsh marigold), or the ability to take up toxic substances (cattails, bulrushes). The iris roots produce a substance that kills *Salmonella* bacteria.

Todd expects the water coming out of the marsh to be as pristine as a mountain stream. He scoops up a beakerful of it. It looks pristine all right. It tests pristine. The plant meets the toughest standards for an advanced treatment system. "Look at these watercress here at the outtake pipe." (They are yellow and shriveled.) "They're *starving*—there's no nutrient left in the water. This stuff is pure!"

The operators over at the breakpoint chlorine plant have difficulty controlling it. "We're here twelve hours a day, watching it like a hawk—it's real operator-intensive." The alkalinity has to be just right, the chemicals have to be in exact relationship to each other. A mistake can produce a chemical excursion; hence the gas masks and warning signs.

The greenhouse is not operator-intensive but organism-intensive. With such a variety of species there are many biological pathways, some of which work on sunny days, some on cloudy, some when it's hot, some when it's cold—as in nature. The worst imaginable mistake might kill off some snails, but it wouldn't require an evacuation.

"I'm amazed at how resilient these creatures are," says Todd. "Even when they're poisoned, they bounce back." Every other morning at 4 A.M. the hot tubs in the sports center are flushed with bromine, and a wave of death flows through the greenhouse, but the populations reestablish themselves. A worse poisoning happened one winter when aluminum from the flocculation system was accidentally routed into the greenhouse inflow pipe. The aluminum tied up phosphate, and many organisms "went on hold." They didn't die, but they didn't grow either. It took six weeks to figure out what was wrong, but again the system recovered.

John Todd figures an acre of greenhouse would be needed for all the sewage of Sugarbush, and 120 acres for the city of Providence, Rhode Island, where he is planning a full-scale plant. "That sounds like a lot of space, but it's about the same as a traditional treatment system. The greenhouses don't all need to be in one place. And they're not spaces you want to stay away from, like regular sewage plants. They're beautiful, safe, productive spaces."

The greenhouses for Providence will not only treat sewage; they will house a business raising flowers, aquarium plants and animals, and medicinal herbs. The capital cost will be about one-third that of an equivalent sewage treatment plant.

Other solar-aquatic plants are being considered for Martha's Vineyard and Santa Rosa, California. A simpler one, not glassed over and using only bacteria and water hyacinth to absorb nutrients, has operated in San Diego for years. It was designed by Steve Serfling of Solar AquaSystems of Encinitas, California. Serfling is a former student of John Todd.

Most of the people involved in the tests at Sugarbush are openminded and interested in all the options. Everyone likes the solaraquatic system because it's cheap, safe, and pleasing. But it's not a system you can set up or shut down quickly. It requires expertise not normal to civil engineers. It's new and strange and unproven. The breakpoint system is expensive and hard to control, but it can go up quickly, and it's understandable to the technically trained minds in the world of sewage treatment.

Concrete block versus arched greenhouse. Pipes and chlorine tanks versus willows and fish. The two systems are almost caricatures of each other and of the engineering and ecological approaches to problem solving. Those approaches are up against each other in every technical arena. Where shall we get our energy? The engineering answer is nuclear power, the ecological answer is solar power. How shall we dispose of garbage? Engineers: mass-burn incinerators. Ecologists: recycling. How shall we raise food? Engineers: agrochemicals and biotech. Ecologists: organic farming.

Maybe we have to make an ultimate choice between these mindsets. Maybe we can and should embrace them both or in some combination. Whatever the best choice is, we're more likely to find it if we set up honest comparisons of real operations, side by side, tested fairly, like the sewage treatment systems at Sugarbush.

DESTABILIZING AND RESTORING PLANETARY SYSTEMS

THE OZONE hole and the greenhouse effect send chills down the backs of environmentalists because these problems are the first signs (that we know of) that human beings are disrupting things on a planetary scale. They are atmospheric phenomena that develop slowly but with a huge momentum. By the time they are obvious, it will be too late to do anything about them. International cooperation will be necessary to correct them. They are problems that make people feel small and hopeless.

They made me feel that way too, at first. But the more I learn about the planet-level megaproblems, the more I see grounds for hope. The solutions to these problems are not exotic; they will not require extreme sacrifices. They are exactly the things we need to do anyway to solve other problems—conserve energy, preserve forests, stop producing toxic chemicals, and, above all, stabilize population growth and channel economic growth toward need not greed. If we have the sense to mend the ozone hole and prevent greenhouse warming, we'll find we've also gone a long way toward correcting local air pollution, acid rain, toxic dumps, energy scarcity, poverty, floods, drought, loss of wildlife, and even the U.S. balance of payments.

There are even some signs, especially in the ozone story, that we might have that sense—all of us, the whole human world acting together in its own defense and for its own good.

The Hole Story

THERE'S BAD news about the ozone layer. A new report shows a loss of stratospheric ozone over the Northern Hemisphere three times larger than scientists had expected.

There's good news about the ozone layer. A far-reaching international agreement limits the production in all major nations of the chemicals that threaten it. Du Pont has announced that it will stop manufacturing those chemicals altogether.

The ozone story has been breaking like that, bad news and good news, for nearly a decade now. One break makes you think the human race has unleashed a chemical disaster that is simply unstoppable. The next makes you think that life on this planet may have a future after all.

Ozone is a rare, unstable gas made of three oxygen atoms stuck together. It breaks up quickly to form plain old two-atom oxygen. About 15 miles up, however, new ozone forms continuously as sunlight hits oxygen. The concentration of ozone up there is only one molecule in 100,000, but that's enough to screen out much of the sun's ultraviolet light.

The UV light that does get through the ozone layer is what gives us sunburns and what harms our eyes when we look directly into the sun. It's a sterilant. When I was a kid, the school shone UV light on our heads to get rid of ringworm. UV light is a packet of energy just the right frequency to break the chemical bonds that hold together living things. Not until the ozone layer formed 500 million years ago could early life creep out onto the land. Until then it had to shelter under the sea—which gives you some idea of what the loss of the ozone layer could mean now.

You've probably heard that ozone depletion will cause human skin cancers. Probably more important, however, could be the harm to human vision and to immune systems. Every form of life exposed to sunshine could be affected in some way, especially green plants. The more the ozone layer is disturbed, the greater the ecological destruction. Quite a price to pay for car air-conditioning and styrofoam hamburger containers!

The cause of stratospheric ozone depletion is a set of man-made chemicals called chlorofluorocarbons (CFCs). CFCs are used in refrigerants, fire extinguishers, plastic foams, aerosol sprays, and a host of industrial processes. The connection between CFCs and ozone was first postulated in 1974 by F. Sherwood Rowland and Mario J. Molina at the University of California at Irvine. They calculated that CFCs could set off an insidious chain of ozone-destroying reactions.

When a CFC molecule reaches the stratosphere, sunlight splits it into, among other things, atoms of free chlorine. The chlorine reacts with ozone to form oxygen and chlorine oxide. The insidious part is that the chlorine oxide then breaks up to form chlorine again, which can bump off another ozone molecule. Each chlorine atom gobbles up ozone but remains itself unchanged, like a gaseous Pac-Man. Scientists estimate that a single atom of chlorine can destroy 100,000 molecules of ozone.

Another insidious fact: A CFC molecule released on the earth's surface can take as long as fifteen years to wend its way up to the stratosphere. The damage we measure now is a result of what we did fifteen years ago—and years' worth of CFC emissions are still on their way up, impossible to stop.

Environmentalists took Rowland and Molina's warning seriously. They agitated until the United States banned the use of CFCs in aerosol sprays in 1979 (they are still legal in Japan and most of Europe). That ban reduced world CFC use by 25 percent. Other uses of CFCs were increasing, however—for coolants in car air-conditioners, for solvents to clean electronic circuit boards, for blowing bubbles in the sort of plastic foam used in hot-drink cups. Now about a million tons of CFCs are produced worldwide each year. Most of them find their way, sooner or later, to the stratosphere.

The United Nations Environment Programme (UNEP) held a meeting in March 1985 to produce the Vienna Convention for the Protection of the Ozone Layer. "That was a dreadful piece of verbiage," says Konrad von Moltke of the Conservation Foundation, who has been in the thick of the ozone battle. "It was an agreement to disagree." The United States, Canada, and Scandinavia, which already had aerosol bans, wanted other countries to adopt them too. Some nations wanted strict limits on all CFC production. Others, with major CFC industries, such as France, West Germany, and Great Britain, were not convinced there was a problem at all.

Just as the Vienna Convention was taking place, the news broke about the "ozone hole" over Antarctica. The British scientific station at Halley Bay had recorded a steady loss of stratospheric ozone above its site since 1977, but the drop was so incredible that the measurements had been suspect. When similar losses were detected over Argentina, the findings were finally believed and published. They were quickly confirmed by satellite data from the Nimbus 7 satellite of the National Aeronautics and Space Administration. The NASA computer had been registering low ozone readings for a decade but discarding them on the assumption that they had to be instrument errors. The readings said that ozone above Antarctica in October, the beginning of the Southern summer, was depleted by 40 to 60 percent.

Discovery of the hole shocked the scientific world and started a burst of research, environmental activism, and diplomatic activity.

The United States broke the ice in the international negotiations in November 1986, when it stopped insisting on an aerosol ban and suggested a 95 percent reduction of all CFC production. Other nations balked at the high percentage but liked the general idea. The World Resources Institute updated European environmental groups and got them involved. German environmentalists, including von Moltke, used a national election to pressure all political parties to advocate a CFC production ban.

A breakthrough came in Geneva in March 1987, when Mustafa Tolba, the director of UNEP, proposed a three-stage plan, called the freeze-20–30 plan. First freeze CFC production at current levels; four years later reduce it by 20 percent; four years after that reduce it by another 30 percent. "No one expected a proposal that strong," said von Moltke, "but no one shot it down."

The idea went back to national governments for consideration. The Reagan administration was thoroughly split on the issue, but the "good guys"—Secretary of State George Shultz and EPA director Lee Thomas—won out, aided by the fact that U.S. manufacturers, such as

Du Pont, were developing CFC substitutes. The freeze-20–30 plan was still alive at a meeting in Brussels in June, and von Moltke was rejoicing. "This goes beyond anything I could have imagined twelve months ago."

In September 1987 in Montreal, twenty-four nations and the European Community formally signed an ozone protocol. The final agreement backed off a bit from the freeze-20–30 plan, but the stipulated CFC reductions amounted to about 40 percent worldwide over ten years.

The Montreal Protocol may be too little, too late. It was greatly strengthened in a subsequent meeting in London in 1990. Third World nations are just beginning to sign it. And signing is no guarantee of action. The pact will have to be enforced. Even if everyone abides perfectly, there are so many CFCs already in the environment and permitted by the agreement that stratospheric chlorine pollution may rise by 25 percent before it begins to turn down. And what's already up in the stratosphere will remain there for 50–100 years.

Though the news keeps breaking, good and bad, the Montreal Protocol is a real achievement. For once, nations have worked together to prevent an environmental catastrophe instead of trying haplessly to repair one that has already happened.

It's worth noting that this great international breakthrough didn't require a stroke of enlightenment causing everyone to become permanently virtuous. There was no world government forcing regulations onto stubborn nations. There was no environmental Gandhi leading the masses into action. What brought about the agreement was dedicated people at many levels, with many skills, playing many roles. For instance:

- Scientists, working through international networks that transcend politics, steadily supplied crucial information to the political process.
- Environmental organizations in many countries mobilized pressure, educated politicians, and achieved early bans on aerosol sprays.
- National governments, in particular those of the United States, Canada, and the Scandinavian countries, led the way in the negotiations.
- Some corporations that make CFCs saw what was coming and went to work to develop alternative chemicals rather than blocking negotiations.

- The United Nations, especially UNEP, provided an international venue and skillful leadership in keeping the discussions going.

A strange assortment of world-savers, often at one another's throats, but all necessary. It's good we have them—we'll need them again.

The Greenhouse Effect, Left, Right, and Center

"HIGH COST of Greenhouse Hysteria," reads one recent headline. Another says, "Greenhouse Effect Looks Like Just a Lot of Hot Air." And a third, "Environmentalists, Not Pollution, Are the Real Threat."

The backlash has appeared. Once the greenhouse effect was popularized, it was sure to become politicized.

Folks who are unhappy with the state of the world anyway are using one hot summer to call for major social reforms. The comfortable people who always oppose change would rather not believe in global warming at all. They will probably go on denying it until they see palm trees grow in New York and rising seas lap the White House steps.

One side likes to quote scientists such as Paul Ehrlich of Stanford, who can always be counted on to be alarming: "We are going to see massive extinction. . . . We could expect to lose all of Florida, Washington, DC, and the Los Angeles basin. . . . We'll be in rising waters with no ark in sight."

The other side searches out Reid Bryson of the University of Wisconsin, one of the few unconcerned climatologists: "The . . . statements that the greenhouse warming is here already and that the globe will be 4 degrees Centigrade warmer in fifty years cannot be accepted."

The earth, of course, does not lean right or left. It has no preference for stability or change in human institutions, nor does it summon authority by selective quotation from biased sources. It operates by inflexible laws, which human beings, when they're not in a knee-jerk political mode, can understand—partially anyway.

Fortunately, as greenhouse politics becomes more muddled, greenhouse science is becoming more clear. I've been watching the literature for twenty years, and I've just come back from one of the innumerable conferences on the subject. There will always be maverick scientists for the politicians to quote, but in fact there is strong consensus on this subject. I'd guess that 99 percent of scientists would agree with the following conclusions.

1. *The phenomenon of greenhouse warming is a certainty.* One of the first Nobel Prize winners, the Swedish chemist Svante Arrhenius, explained 100 years ago how greenhouse gases trap the sun's energy and warm the earth. He warned that one of those gases—carbon dioxide—is released when we burn oil, gas, or coal, and he calculated how much the earth would warm if we burned enough fuel to double the carbon dioxide in the atmosphere.

A hundred years later his reasoning is accepted, and his calculations are still close to the mark. The main thing that has changed since Arrhenius's time is that humans now create gases he never heard of, such as chlorofluorocarbons (CFCs), which are even more powerful and long-lasting heat traps than carbon dioxide (and which, in the case of CFCs, also destroy the ozone layer).

2. *Greenhouse gases in the atmosphere are increasing faster and faster.* They have been measured directly for thirty years and indirectly (from air bubbles buried in polar ice) for 160,000 years. Different gases are increasing at different rates, but all of them in upward-rising curves. More are added to the atmosphere each year than the year before.

If these atmospheric changes go on at their current rates, the equivalent of Arrhenius's carbon dioxide doubling will be reached about the year 2030—when a child born this year is middle-aged. That may not happen because we may stop it. It also could continue well beyond a mere doubling if we make no changes.

3. *We are the cause of the increase.* We cause it primarily by fuel burning, air pollution, the manufacture of CFCs, and deforestation.

Figure 5

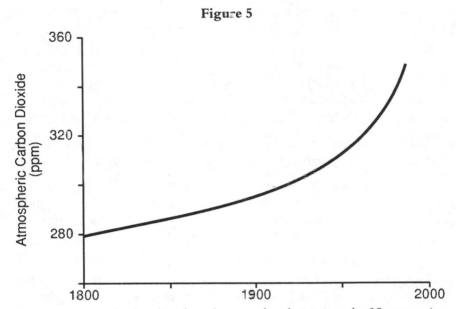

The concentration of carbon dioxide in the atmosphere has gone up by 35 percent since pre-industrial times, and it's still rising—exponentially.

SOURCE: UNEP

These activities are also increasing faster and faster with the growth of the human population and economy. (There is a global agreement to reduce CFC production, which hasn't yet gone into effect.)

4. There is a long time lag between greenhouse gas increase and measurable climate change. That's partly because the oceans and ice sheets act as a delay or drag. It's partly because climate is a lagged indicator—climate is the average of weather over one or two decades. Weather is variable, noisy, shifting. You have to keep track of it for years before you can be sure the climate has changed.

Therefore, if we wait to prove the greenhouse effect by measuring climate change, we will confirm it several decades too late to do anything about it. We should take our cues from the leading indicators—the atmospheric concentrations of greenhouse gases. Their message is undeniable.

5. The exact nature of a greenhouse climate change is highly uncertain. Anyone who tells you that Russia will be better off with a global warming, or that the Midwest will dry up, or that there will be more rain in the Sahara is going well beyond what scientists know. Just which places on the planet will be hotter, colder, wetter, or drier is still a matter of controversy. (The scoffers like to "disprove" the greenhouse effect by showing that a single place—such as the continental United States—has cooled lately. Those data, derived from just a small percentage of the earth's surface, are irrelevant to the global picture.)

There are two other messages of particular importance to politicians that are emerging strongly from the scientific meetings.

6. *A climate change as rapid as the greenhouse change is likely to be will be disastrous for the earth's ecosystems and for the human economy.* We are more dependent than we realize on the seacoasts, rainfalls, rivers, seasons, plants, and animals occurring just the way they occur now.

7. *Only a small amount of global warming is now inevitable.* The greenhouse effect is talked about too much as a matter of destiny and too little as a matter of choice. The greenhouse gases already in the atmosphere will create a global warming, scientists think, of only 1 or 2 degrees. The real danger is the gases we will put in if we go on accelerating at our current pace.

We Can't Afford the Greenhouse Effect

FORWARD-THINKING PEOPLE are preparing for the greenhouse effect. Barge companies on the Mississippi River are acquiring railroads in case the river becomes permanently unnavigable. Planners of Boston's sewage treatment system are taking into account a sea-level rise from global warming. The Weyerhaeuser Company is planting drought-resistant trees. The Dutch are raising their dikes.

Either this behavior is certifiably crazy, or it's an apt assessment of the craziness of the human race as a whole—a reasoned bet that we will be stupid enough to let the greenhouse effect happen.

I'm not that much of a pessimist. I think we're rational animals, at least when it comes to economics. The economics of the greenhouse effect are quite clear. We can't afford it.

The Environmental Protection Agency has commissioned studies of the effects on the United States of a projected global climate change. In unemotional bureaucratic language, the report spells out unmitigated disaster.

> Wheat and corn production may . . . shift away from the Great Plains. . . . The agricultural economy may no longer be able to sustain the rural population.
>
> Under the driest scenario, projections for the Great Lakes region and New England are that species like eastern hemlock and sugar maple could disappear. Mature natural forests in the region could be reduced from one-quarter to one-half their present volumes, . . . with many poor sites . . . giving way to grassland or scrub conditions. . . . There will probably be disruptions and/or reductions in the availability of major forest resources—wood, water, wildlife, recreation opportunities.
>
> The United States could lose 30 to 70 percent of its coastal wetland with a one-meter rise in sea level. . . . A one-meter rise would inundate an area the size of Massachusetts. Most of these losses would be concentrated in the Southeast particularly Louisiana and Florida.

The EPA says that coastal areas now in 100-year floodplains would in fact be subject to storm surges on average every fifteen years. Hurricanes would form more often and be stronger. Salt water would invade groundwater aquifers, and the salt-fresh interface would move higher up river mouths, endangering water supplies from Cape Cod to New York to Miami to California's Central Valley.

Protecting coastal cities with bulkheads, levees, and pumping systems would cost $30–100 billion. Raising barrier islands by pumping sand onto them would cost $50–100 billion. (That would double property tax to residents of those islands, says EPA, but that's better than losing their property altogether.) "We will probably have to gradually remove structures from much of our coastal lowlands," says the agency. "Although this will probably not be necessary for several decades, we need to lay the groundwork today."

The warming could increase the need for new power plants by 14 to 23 percent; on average, that means four to eight new 1,000-megawatt power plants *in each state*. That's 200 to 400 plants. At, say, $2 billion apiece, it would cost $400–800 billion for construction, not counting fuel and operating costs or the fact that, with rivers and seacoasts moving around, cooling water could be hard to find.

The Great Lakes are expected to fall, shipping channels will have to be dredged, pollution will be less diluted, algae growth will increase, fish will die. Mountain snowmelts will be greater and come earlier, requiring flood protection downslope and leaving less runoff for the summer when it's most needed. As precipitation patterns change, virtually every reservoir and dam in the nation will become either too large or too small.

Iowa's corn depends not only on Midwest temperature and rainfall but also on Iowa's rich prairie loam. If good corn weather shifts north, not only will it cross into Canada, it will move onto thin, glaciated northern soils. There might not be a cornbelt at all.

Forests, prairies, and wetlands will not simply pick up and move. Every kind of bird, bug, tree and soil mite will respond in its own way and at its own pace to temperature and moisture change. Pests may move faster than controlling predators, and animals may move beyond sheltering habitat. Ecosystems will be taken apart and will have to come together in new combinations. Ecologists expect pest outbreaks and extinctions. Adaptable species like English sparrows, deer, coyotes, raccoons, and rodents will probably thrive.

Those are the possible effects on just one nation. Similar changes will be happening all over the world.

We could grin and bear these changes, spend trillions of dollars trying to adjust, and still absorb trillions of dollars worth of losses, or we could allocate a small fraction of those trillions up front to prevent the global heat trap from forming in the first place. Only a small amount of global warming is now inevitable. That amount increases every day we go on dithering.

Avoiding the Greenhouse and Making a Better World

BY SOME stroke of providence, every step we need to take to reverse the greenhouse effect is worth doing anyway. Every step is technically possible. Most of them will even save money. Unfortunately for our sense of drama, preventing global climate change need not be a great, grim sacrifice. We are not called upon to save the world. Instead, we have an opportunity to build a better one.

Here's a list of antigreenhouse measures, in order of effectiveness, as I heard them presented to an international meeting recently:

- Use energy much more efficiently (which would also cut fuel bills, urban smog, acid rain, oil spills, toxic wastes, and oil imports).
- Phase out chlorofluorocarbons (CFCs) completely and quickly (which will also help repair the ozone hole).
- Accelerate the transition to solar, wind, hydro, and biomass energy sources (which would have all the beneficial effects of energy efficiency and ensure an inexhaustible energy supply).
- Shift fossil fuel use away from coal and oil and toward natural gas (an interim measure until solar sources are tapped—it would also reduce many air pollutants).
- Stop deforestation and accelerate reforestation (thereby sustaining the supply of forest products, reducing soil erosion, flood, and drought, moderating temperatures, and preserving endangered species).
- Increase use-efficiency and recycling of all materials and of water (which would save money, reduce energy needs, extend the lives of mines and groundwaters, and reduce municipal solid waste, toxic waste, mine waste, and water pollution).

- Practice low-input agriculture (reducing farm costs, increasing energy efficiency, restoring soils and wildlife, reducing water pollution, and improving health).
- Slow population growth in poor countries, where 90 percent of population growth takes place. Slow wasteful consumption in rich countries. (These two measures would ease every environmental problem and most economic ones—and without them the other steps listed here would just be stopgaps.)

Some people find this list an exciting challenge. Others think it sounds like Exercise Properly, Get Enough Sleep, Floss Your Teeth. We know our lives would be better if we did these things, but there's a huge barrier of habit to get over—we'd really rather not think about it.

People will never change their comfortable self-destructive habits, some say. Let's talk about adapting to climate change; we'll never prevent it.

I'm not willing to be that fatalistic. I believe in good old self-serving human rationality. I think that anyone, however lazy or greedy, who looks at the full costs and benefits of preventing climate change, as opposed to enduring it, will see that there is no better payoff on the planet than preventing greenhouse warming.

At least the first few steps could be trivially easy. If West Germany put a speed limit on its Autobahns, it would reduce its greenhouse gas emissions by 26 million tons per year. By buying efficient refrigerators, Americans have already cut electricity use enough to avoid building eighty coal-fired power plants (and saved $5 for every $1 spent on the refrigerators). We could save more money and reduce refrigeration energy by another two-thirds by installing the most efficient models now on the market.

Who should lead the way? Governments? Individuals? The only possible answer is both, with the realization that in democracies the people are always out in front. And there's plenty we can do. The Greenhouse Crisis Foundation has come out with a list of 101 things you can do to stop global climate change. You can imagine what's on it.

Buy energy-efficient appliances and a car with the highest possible gas mileage. Insulate your home, caulk and weatherstrip your doors and windows, turn off lights when you're not using them. On nice days use the sun, not the clothes dryer. Reuse, repair, and recycle everything you can. Don't buy stuff you don't need. Shop with a reusable canvas bag; turn down both paper and plastic bags from the stores. Shun overpackaged products and complain to manufacturers about them. Don't take unnecessary car trips. Don't speed. Buy organic food, not junk food. Plant trees. Plant a garden, and don't use chemical fertilizers or pesticides on it or your lawn. Write to Congress and the president.

It may sound like Eat Balanced Meals, Don't Smoke, and Balance Your Checkbook, and in a way it is—it's what adult people need to do, as an unremarkable matter of habit, to make their lives and their planet work. As Winston Churchill once said, "Sometimes you have to do what is required."

Or, paraphrasing Buckminster Fuller, it's time to grow up. The human race has been like a bird in the egg, supplied with an abundance of nutrient and an unpolluted space into which to develop—to a certain point. We've exhausted the nutrient and the space. "We are going to have to spread our wings of intellect and fly, or perish."

Burning Coal in Connecticut, Planting Trees in Guatemala

APPLIED ENERGY Services (AES) of Arlington, Virginia, has come up with a way to do something that everyone thought was impossible. It is building a coal-burning electric plant that will not add carbon dioxide to the atmosphere—which means it will not worsen the greenhouse effect.

The plant will also be energy-efficient. Its air pollution emissions will be low. Indirectly, it will help combat soil erosion and alleviate poverty. Sounds unbelievable, but it's all true.

The founder and president of AES, Roger Sant, was head of energy conservation for the Federal Energy Administration (before there was a Department of Energy) under President Ford. When Sant left government, he set out to help industry save energy. One way to do that is with cogeneration.

Cogeneration means using the "waste heat" from an electric power plant for a purpose more productive than warming up a river, such as heating a building or providing steam for an industrial process. The plant AES is constructing in Uncasville, Connecticut, will sell electricity to Connecticut Light and Power and steam to the Stone Container Corporation.

It is a "clean coal" plant, using a technique called fluidized bed combustion. Instead of a scrubber in the stack to take out the sulfur dioxide that causes acid rain, limestone is mixed with the coal right in the furnace. The sulfur dioxide is absorbed before it ever gets to the stack. The furnace also burns at a low enough temperature to reduce formation of nitrogen oxide, another cause of acid rain.

The main gas coming out of an AES smokestack is carbon dioxide, the inevitable product of burning the carbon in coal and until recently not considered a pollutant. But Roger Sant knew that carbon dioxide is a greenhouse gas. He kept going to meetings about the greenhouse effect. He'd come back to his company and say, "This greenhouse thing is really bad. We have to do something."

He assigned his strategic planning team, headed by Sheryl Sturges, to figure out what to do. The team came up with several options:

1. Go into some other business.
2. Capture the carbon dioxide and inject it into oil wells, where it could enhance the recovery of oil and stay underground where it won't cause greenhouse warming.
3. Capture the carbon dioxide and sell it to soft drink companies to make bubbles.
4. Plant trees, which will fix carbon dioxide into wood.
   "How many trees?" asked Sant.
   "About 52 million," said Sturges.

Enter Paul Faeth of the International Institute for Environment and Development, who is an expert not on power plants but on trees. AES hired him to select through an open-bidding process some agency, anywhere in the world, that could assure the planting and maintenance of lots and lots of trees. It doesn't really matter where the trees are. The Uncasville plant puts carbon dioxide into the global atmosphere, and trees anywhere can take it out again.

The winning bid came from a coalition of CARE, the Guatemalan government, the Peace Corps, and USAID. They had an ongoing tree-planting program involving 40,000 small-scale farmers. For $2 million from AES, they could extend the program by 194,000 acres—52 million trees.

Half of Guatemala's forests have disappeared over the past thirty-five years. The result is disastrous erosion and flooding. The CARE program helps farmers plant trees at the edges of fields and on eroded hillsides. The trees are used for fruit crops, lumber, fuelwood, living fences, building poles, and, very importantly holding soil and water. The program is wildly popular—farmers are demanding five times more seedlings than the tree nurseries can supply.

AES staff who have visited Guatemala talk not so much about fixing carbon from the coal plant as about the need for trees. Dave McMillen, the no-nonsense manager of the Uncasville plant, admits that he thought it was crazy that his budget and oversight responsibility should include forests thousands of miles away. But when he went to Guatemala, he became, as his colleagues say, "Mr. Tree." He could hardly believe the poverty of the people, the farmers hanging from ropes to hoe their steep garden plots, the denuded land, the silt running off the hills and filling up the reservoirs behind hydropower dams—and the good that trees can do for the people and land of Guatemala.

Paul Faeth says, "The carbon fixation is nice, but the social benefits are even greater. There will be fuelwood, better soil, higher crop yields, jobs. If I never do anything else in my career, I feel I've helped do a little good here."

For its next power plant AES is looking for a way to plant 100 million trees.

Has AES hit upon a scheme to go on burning all the coal we'd like while avoiding the greenhouse effect? No, says Roger Sant, it's only a way to buy time. "We're going to have to stop burning fossil fuels someday. We have to figure out how to invest $250 million not in a power plant but in energy conservation. That's the next step I'd like to take."

Biosphere I and Biosphere II

O UT IN the Arizona desert they're building Biosphere II, the first terrarium to contain human beings. Enclosed under 2.25 acres of glass will be a rain forest, a 35-foot-deep ocean, a marsh, a desert, a half-acre farm, goats, chickens, and fish. Eight scientists will live there for two years. Nothing will go in, nothing come out, except electronic messages and energy from the sun.

In other words, Biosphere II will run on the same principles as Biosphere I, the earth itself.

Biosphere II is a commercial venture, intended to spin off inventions and crop varieties useful in space and also on earth. It will have to purify air and water, recycle sewage and solid waste, and grow food— all without chemicals—and these technologies may be marketable. But I suspect the biggest payoff will be appreciation for Biosphere I.

At least that has been the experience of the two biosphere designers I know, Vladimir Iakimets, who is Russian, and John Todd, who is Canadian.

Iakimets was one of the designers of a Soviet space station intended to support human beings for two to three years. He was a mathematician when he started, with little knowledge of biology, but as he reasoned the problem out, he learned, step by step, how a living planet works. There have to be green plants to turn the carbon dioxide expired by people back into oxygen. For food there have to be higher plants. To make use of all parts of the plants, there have to be animals.

To return animal wastes to nutrients to grow more plants, there have to be microbes and soil.

Pretty soon Iakimets found he needed to lift just about everything on earth up into space. Unlike the designers of Biosphere II, he couldn't even count on gravity. If he wanted water to flow somewhere, he had to pump it. And without gravity, plants might not even know which direction to send their roots.

To test that out, Russian cosmonauts tried growing plants on their long space voyages. Tulips didn't do well in space, Iakimets told me, but onions did. The cosmonauts grew onions, harvested one at each stage of growth, and stored it in alcohol to be studied back on earth. Except on the last day. The cosmonauts had had enough of gooey space food squeezed out of tubes. They snipped the bloom off the onion, put it in the alcohol, and ate the rest.

Iakimets loves that story because it reinforces his opinion that human beings will need to take into space a great variety of their living companions on earth, for psychological happiness as well as biological maintenance. He quotes the motto of the Soviet space program, "On Mars we will have to grow apples." That means, says Iakimets, that the human body and spirit are so tied to earthly ecosystems that we cannot leave those ecosystems; we have to take them with us.

While he was making his calculations, Iakimets was falling in love with his planet. The more he learned, the more he was impressed with the interconnected systems of the earth's biosphere. He saw how difficult and expensive it would be to duplicate even a fraction of them, how much simpler it is to protect this earth than to create a new one. Out of the process he became a raging environmentalist—one of the few Soviet environmentalists I know.

At the same time on the other side of the world John Todd was building bioshelters. Still standing at the New Alchemy Institute on Cape Cod are many of the structures Todd created over the past fifteen years, all of them little experiments of the Biosphere II sort. Algae grow in solar-heated cylinders of water, fish eat the algae, and the wastes accumulating at the bottoms of the tanks become fertilizer for greenhouse plants. Beneficial insects are introduced to eat up harmful insects. Everything is heated by the sun; water is pumped by windmills. Todd was fooling around with these systems not to go into space but to figure out how to live a better life on earth.

As the structures got bigger and their biological communities more complex, I remember Todd reflecting about the use of his technologies in space. "Too many times I've come out and found my whole system dead," he told me once, "because of a toxin or a bacterium I didn't know was there. And sometimes I've been bailed out of a difficulty because some complex of organisms turns up that recycles something or eats some waste product. I didn't put them there; sometimes I didn't even know they existed. We are still too ignorant about this wonderful planet to duplicate its functions in space."

In other words, Todd is a raging environmentalist too. He operates with deep respect for natural systems, a respect that grows the more he learns.

Todd is not personally involved in the Biosphere II project, but he is excited by it as an educational tool and a research project. His eyes sparkle when he talks about it. "By putting people in it, they're playing real hardball. Mistakes will be life-threatening. Think how much they'll learn!"

Maybe they'll learn, maybe we'll all learn, as Todd and Iakimets have, to treasure Biosphere I, which nurtures us, purifies our air and water, and does its best to process our ever-mounting piles of waste. Biosphere II even gives us the first rough estimate of what those free services are worth. The project in Arizona has cost roughly $100 million to build. By extrapolation that makes the surface of Biosphere I worth $1,500 million billion.

A VOICE
CRYING FOR THE
WILDERNESS

THE MOTTO of Dartmouth College is *Vox Clamantis in Deserto*—a Voice Crying in the Wilderness—emphasizing, I guess, the time when the college was founded, when it was the only institution resembling a school in all the forested miles around it. Dartmouth is now surrounded by shopping malls and stylish condos. I wish these days that the motto, and the reality, could be changed to *Vox Clamantis PRO Deserto*—a Voice Crying *FOR* the Wilderness.

I wish it could be the motto of the human race.

Next to destabilizing the atmosphere, the most profound and unforgivable thing humans are doing is eliminating the other species, the natural wealth, the biodiversity of the planet. I notice, as I read over the columns I've written on this subject, that they are nearly all mournful. I can see solutions to energy problems, food problems, and waste problems, which require of human society only rationality—not a paradigm change or a newfound morality. In the area of respect for wilderness and wild creatures, however, I see the necessity of an entirely new worldview, one in which *Homo sapiens* no longer figures as the predominant species on the planet.

There are signs of that worldview. Something in everyone responds to the mystery and beauty of the planet's biological systems. But in our everyday actions we are a long way from Aldo Leopold's famous land ethic: "A thing is right when it tends to preserve the integrity, stability, and beauty of the biotic community. It is wrong when it tends otherwise."

Every day, everywhere on earth, we are desecrating the integrity, stability, and beauty of nature. Something deeply sacred is being irrevocably lost. I can only lament.

How Much Abuse Can the
Earth Take?

A PESTICIDE PLANT turns the Rhine into a "dead river" the entire length of Germany. Acid rain falls on Scandinavia. Oil spills destroy the coral reefs of the Red Sea. Air pollution kills vegetation in the Los Angeles basin. Radioactive waste dumped into the Irish Sea can be traced up to the Arctic Circle and over to Denmark.

We think of these problems as separate, and we try to deal with them one at a time. But earth's ecosystems are not separate. They are woven together by flows of energy, water, nutrients, and air into one fabric, which links all living things and which makes all life possible.

Even the most tightly woven fabric will fall apart if it has enough holes poked in it. Are we close to that point? How much abuse can the planet take?

In 1986 an assemblage of field ecologists got together in Woods Hole on Cape Cod to address that question. They are people who study the creatures that live in a lake or a forest, what eats what, how the beautiful, intricate machinery of nature fits together. They are experts on the Arctic or the peat bog or the prairie. Together they form a community of monitors, documenting what is happening to the living systems of the earth. They told some harrowing stories.

Lichens are like the canaries that warned miners of bad air quality. An undisturbed lichen can live for hundreds of years, but in polluted air it dies. Parts of England have already lost 89 percent of their lichens. Southern California has lost 50 percent. The "lichen desert" around the city of Zurich has increased in area by a factor of nine since 1936.

Satellite photos of Amazonian Brazil show the expanding grid pat-

215

tern of settlers' clearings. The pastures that replace the forests erode so rapidly that after twelve years their productivity drops by half. No forest can grow there again.

As the western United States is overgrazed, the native bunchgrasses decline, and foreign cheatgrass moves in. Cheatgrass, unlike bunchgrass, dries in summer, which makes it poor for grazing and a fire hazard. The fires destroy juniper and pine forests on the uplands, opening more room for cheatgrass. Thousands of square miles of the West have been converted from bunchgrass, sagebrush, pine, deer, and elk to cheatgrass, cheatgrass, and more cheatgrass.

Most damaged ecosystems are not lifeless. But the life forms are changed and simplified. Sedges, grasses, and insects are survivors. Trees, mammals, and birds are not.

The changes can be irreversible. The 100-mile swath of devastation downwind from the International Nickel Refinery in Sudbury, Ontario, has not recovered, though pollution emissions dropped by a factor of five after 1972. On the fringe of the damage zone red maple roots are sending up shoots again, but the shoots do not survive. In some places so much vegetation has been lost that the soil has washed away. Once there were farms and forests; now there is nothing but bare rock.

A lake in Canada was deliberately acidified to study the effects of acid rain. As the pH went down, species after species disappeared. When the lake was restored to normal, many did not return. An acid-resistant stickleback that had never been seen in the lake took the place of the fathead minnow. The lake trout, which had nearly starved, grew fat again but did not breed.

Environmental destruction is usually justified by economic arguments. Robert Repetto of the World Resources Institute challenges those arguments. For example, the companies clearing the Amazon forests are realizing about 250 percent on their investment, which explains very neatly why they are there. But the Brazilian government is subsidizing their logging through road building, tax reduction, and other incentives. Counting the whole cost, including that to Brazilian taxpayers, the enterprise is losing 55 percent per year!

As the 1986 Woods Hole meeting went on, the stories the field ecologists told began to add up to a clear message. From tundra to rain forest to ocean depths there is piecemeal degradation. No one knows how many such holes the fabric can sustain. And some human activities, such as fossil-fuel burning, are now so pervasive that they are changing the atmosphere and the climate. For the first time the damage is not only local but global.

I came out of the conference filled with a peculiar mixture of excitement, resolve, and dread.

Surely, I thought, if the ecologists' message is delivered to the world's people, we will all insist that the damage be halted. We have been endowed with logic; we know it is irrational to destroy our resource base. And we also have a moral sense that tells us that our planet is a temple, full of magnificent communities of life that we do not understand, that we did not create, and that we have no right to destroy.

All that needs to be done is to communicate the message, and then we will wake up and start managing ourselves and the planet wisely.

Right?

What Is Biodiversity and Why Should We Care About It?

Most of us have grasped the idea that there's a hole in the sky over the South Pole that could give us skin cancer. We are beginning to understand that a global warming could inundate Miami Beach and make New York even more unbearable in the summer. There is another environmental problem, however, that doesn't have a catchy name like "ozone hole" or "greenhouse effect," and that hasn't yet entered the public consciousness. It's the loss of biodiversity.

Bio-*what*?

Biodiversity sounds like it has to do with pandas and tigers and tropical rain forests. It does, but it's bigger than those, bigger than a single species or even a single ecosystem. It's the whole, all of life, the microscopic creepy-crawlies as well as the elephants and condors. It's all the habitats, beautiful or not, that support life—the tundra, prairie, and swamp as well as the tropical forest.

Why care about tundras and swamps? There's one good reason—self-interest. Preserving biodiversity is not something to do out of the kindness of our hearts, to express our fondness for fuzzy creatures on Sunday mornings when we happen to feel virtuous. It's something to do to maintain the many forms of life we eat and use, and to maintain ourselves.

How would you like the job of pollinating all trillion or so apple blossoms in the state of New York some sunny afternoon in late May? It's conceivable, maybe, that you could invent a machine to do it, but inconceivable that the machine could work as efficiently, elegantly, and cheaply as honeybees, much less make honey.

Suppose you were assigned to turn every bit of dead organic matter—from fallen leaves to urban garbage to road kills—into nutrients that feed new life. Even if you knew how, what would it cost? Uncountable numbers of bacteria, molds, mites, and worms do it for free. If they ever stopped, all life would stop. We would not last long if green plants stopped turning our exhaled carbon dioxide back into oxygen. The plants would not last long if a few beneficent kinds of soil bacteria stopped turning nitrogen from the air into fertilizer.

Human reckoning cannot put a value on the services performed for us by the millions of species of life on earth. In addition to pollination and recycling, these services include flood control, drought prevention, pest control, temperature regulation, and maintenance of the world's most valuable library, the genes of all living organisms, a library we are just learning to read.

Another thing we are just learning is that both the genetic library and the ecosystem's services depend on the integrity of the entire biological world. All species fit together in an intricate, interdependent, self-sustaining whole. Rips in the biological fabric tend to run. Gaps cause things to fall apart in unexpected ways.

For example, attempts to replant acacia trees in the Sahel at the edge of the Sahara desert have failed because the degraded soil has lost a bacterium called rhizobium, without which acacia trees can't grow. Songbirds that eat summer insects in North America are declining because of deforestation in their Central American wintering grounds. European forests are more vulnerable to acid rain than American forests because they are human-managed, single-species plantations rather than natural mixtures of many species forming an interknit, resilient system.

Biodiversity cannot be maintained by protecting a few charismatic megafauna in a zoo, nor by preserving a few greenbelts or even large national parks. Biodiversity can maintain itself, however, without human attention or expense, without zookeepers, park rangers, foresters, or refrigerated gene banks. All it needs is to be left alone.

It is not being left alone, of course, which is why biological impoverishment has become a problem of global dimensions. There is hardly a place left on earth where people do not log, pave, spray, drain, flood, graze, fish, plow, burn, drill, spill, or dump.

Ecologists estimate that human beings usurp, directly or indirectly, about 40 percent of each year's total biological production (and our population is on its way to another doubling in forty years). There is no biome, with the possible exception of the deep ocean, that we are not degrading. In poor countries biodiversity is being nickeled and dimed to death; in rich countries it is being billion-dollared to death.

To provide their priceless service to us, the honeybees ask only that we stop saturating the landscape with poisons, stop paving the meadows and verges where bee food grows, and leave them enough honey to get through the winter.

To maintain our planet and our lives, the other species have similar requests, all of which, summed up, are: Control yourselves. Control your numbers. Control your greed. See yourselves as what you are, part of an interdependent biological community, the most intelligent part, though you don't often act that way. Act that way. Do so either out of a moral respect for something wonderful that you did not create and do not understand or out of a practical interest in your own survival.

The Tragedy of the Commons on Georges Bank and Elsewhere

I N THE Massachusetts State Capitol there is a wooden statue of the sacred cod, a tribute to the massive fishing ground called Georges Bank. For 200 years codfish from Georges Bank have enriched New England. Now, says the Northeast Fisheries Center, the cod population of Georges Bank is collapsing.

This unnecessary tragedy-in-the-making is directly parallel to other unnecessary tragedies, such as ozone holes and national debts, greenhouse effects and urban air pollution. All of them are examples of the "tragedy of the commons."

Every environmental science course teaches the tragedy of the commons, first described by biologist Garrett Hardin in 1968. The commons was the common grazing area in the middle of a village, where everyone was free to pasture a cow. Since the grazing was free, it was to everyone's interest to graze two cows. Or three. If there were no constraints, soon there would be so many cows that the grass would be destroyed. Then the commons would support no cows at all.

In the long term, of course, it makes sense to limit the cows and preserve the commons, the solution that Hardin calls "mutual coercion, mutually agreed upon." It's easy enough to see with cows.

It's easy to see with fish, too. A Norwegian student of mine rushed home to tell his father, a commercial fisherman, about the impending tragedy of the commons off the shores of Norway. "Of course," replied the father calmly. "In a few years there will be no more fish. We all know that. But by then I will have paid off my boat."

Everyone knows. But everyone is rewarded in the short term for overloading the commons, and no one likes mutual coercion, mutually agreed upon.

Two out of every three cod in Georges Bank perished in 1987, mostly in the nets of fishermen. The catch was the highest on record, and it took a record number of fishing days to harvest it. Surveys show few young fish coming along to restore the population.

The Northeast Fisheries Center says there is only one way to prevent the cod from going the way of the haddock (which crashed in the early 1980s and has not recovered) or the way of the thirteen other major world fisheries (out of nineteen) that have collapsed. The solution is to cut the catch in half until the population can restore itself. Either the fishermen cut back now, or nature will force a cutback in another few years.

The director of the New England Fisheries Management Council—which was set up to protect Georges Bank after the United States got rid of international competitors by declaring a 200-mile limit—refuses to consider catch limits. He says the industry's current fish size and mesh-size limits will suffice. "There isn't any doubt that stocks are in bad condition," he admits. "Nobody denies that." But he doesn't believe the fishery will collapse.

In environmental science classes we play a simulation game, in which students manage competing fishing fleets. They are charged realistically for boats, for fishing expenses, for keeping boats idle in port. They're paid for the fish they catch. A small computer program calculates the reproduction and growth of the fish population. The players don't know the actual fish numbers, any more than real fishermen do, except by the evidence of their catch. The players inevitably destroy the fishery, even after they have discussed the tragedy of the commons.

The largest catches, the greatest number of boats, the longest fishing hours always come just as the fish population begins to turn down. The players are usually arguing about when the crash will occur as they send out the boats that will cause it. After the inevitable round of bankruptcies, the depleted fishery supports only one-tenth to one-third as many boats as it could have, sustainably, forever, if it had not been overfished.

If you wonder how fishermen could be so stupid, consider the willingness of industry to limit emissions that cause the pollution of common air and the acidification of common rain. Consider the fact that Americans have voted in three consecutive elections not to be taxed, but to spend for common purposes and let debt pile up—

meanwhile discussing endlessly when the crash will come. Watch communities refuse to put limits on developments that destroy the common amenities that attract the developments in the first place.

How would you feel about mutual coercion, mutually agreed upon, if it required you to drive a higher mileage car or take public transportation, insulate your house properly, and in other ways stop pouring carbon dioxide into the atmospheric commons and deranging the climate of the earth?

What, limits? Requirements? Inconvenience and expense and infringement on my freedom? Why me? I'm just a small part of the problem. I have enough to worry about just making ends meet.

That's it. That's the logic of the cod fishermen, the logic of all the rational, reasonable perpetrators of the tragedy of the commons.

The problem of the mismanaged commons is not hard to solve. The solution doesn't demand that we be angels. It just asks us to lift our heads from the hot pursuit of immediate gratification, lean back, look at the workings of the whole system, and sacrifice some short-term self-interest for long-term self-interest.

Going, Going, Gone

HOW MUCH of the earth's surface would you guess is still wilderness?

According to a Sierra Club survey, about one-third of the earth's land area can still be called wilderness (defined as at least 1 million acres upon which there is no road, powerline, airport, dam, or other major human construction). Over 60 percent of this remaining wilderness is tundra, ice, or desert. Only 12 percent lies in the Tropics.

By the Sierra Club's definition of wilderness, just two areas of wilderness remain in the United States (outside Alaska)—the Greater

Yellowstone area and the Bob Marshall Wilderness in Montana. In Europe, India, and most of China there is no wilderness.

Information like that usually evokes two kinds of reactions: One is, So what? The other is, Let's save every bit we can, while there's still time!

Says Justin Dart, an adviser to former President Reagan: "I am for preservation. I say we should preserve the redwoods, sure, maybe 100 acres of them, just like the way God intended them, to show the kids. Those environmentalists who talk about preserving the wilderness in Alaska—how many goddamned bloody people will end up going there in the next 100 years to suck their thumbs and write poetry?"

Said Henry David Thoreau: "At the same time that we are earnest to explore and learn all things, we require that all things be mysterious and unexplorable, that land and sea be infinitely wild, unsurveyed and unfathomed by us. . . . We need to witness our own limits transgressed, and some life pasturing freely where we never wander."

Wilderness, said President Theodore Roosevelt, must yield to the needs of people: "Surely . . . a rich and fertile land cannot be permitted to remain idle, to lie as a tenantless wilderness, while there are such teeming swarms of human beings in the overcrowded, overpopulated countries."

Says Paul Gruchow in his book *The Necessity of Empty Places* (1988): "Empty; unoccupied or uninhabited; unfrequented. . . . Empty is one of those words that reveals unspoken attitudes. Lacking people, it means. No humans equals nothing. . . . The word 'empty' inherently expresses contempt for everything that is not human. The old puzzle about the tree falling in an unoccupied forest would not be a puzzle at all in a world where trees and porcupines, say, were assumed to have some justification independent of humanity."

Wilderness is stupid, wilderness is sacred, wilderness is a resource for human needs, wilderness lifts us from our limited self-centeredness. In the paralysis engendered by these opposing views, one other view briskly prevails: Wilderness is a place to make money.

Ecologist Paul Ehrlich once asked a Japanese journalist why the Japanese whaling industry is busily exterminating the very source of its wealth. The answer: "You are thinking of the whaling industry as an organization interested in maintaining whales. Actually it is better viewed as a huge quantity of capital attempting to earn the highest possible return. If it can exterminate whales in ten years and make 15 percent profit, but it could only make 10 percent with a sustainable harvest, then it will exterminate them in ten years. After that, the money will be moved to exterminate some other resource."

The same logic propels the burning of the Amazon rain forest, the search for oil in the Arctic, the cutting of old-growth forests, the filling in of wetlands, the commercialization of the world's most beautiful places. Left to its own momentum, that logic will generate mountains of money and no wilderness.

About 8 percent of the remaining wilderness (just over 2% of the earth's surface) has been designated, on paper at least, as national park or nature preserve. UNESCO is engaged in a worldwide campaign to establish at least one Biosphere Reserve in each of the 200 "biotic provinces" it has identified on earth. So far only one-third of the needed reserves have been established. The cost to complete the job and manage all the reserves would be about $150 million a year—about 75 minutes' worth of the world's military expenditures.

Which of the many human opinions about wilderness will dominate over the next fifty years? The answer matters greatly. It will determine whether there will be any wilderness left at all.

Kentucky farmer and writer Wendell Berry: "We have never known what we were doing because we have never known what we were undoing. We cannot know what we are doing until we know what nature would be doing if we were doing nothing."

Edward Hoagland: "The swan song sounded by the wilderness grows ever fainter, ever more constricted, until only sharp ears can catch it at all. It fades to a nearly inaudible level, and yet there never is going to be any one time when we can say right NOW it is gone."

Henry David Thoreau: "In Wildness is the preservation of the world."

Wendell Berry: "In human culture is the preservation of wilderness."

Fun with an Endangered Species

I COULDN'T BELIEVE it when I saw the item on the town meeting agenda: "Article 26. To see if the town will vote to designate the cobblestone tiger beetle, an endangered species that inhabits Plainfield's own Burnap Island, the Plainfield Town Insect."

"Some environmentalist has gone off the deep end," I thought, uncharitably. I could imagine the maudlin appeal being prepared somewhere in town. You know the type. Serious and superior. The sacredness of nature's web. The importance of genetic diversity. The crass commercialism of the entrepreneurs who propose hydropower dams on the Connecticut River, dams that would flood Burnap Island and doom the tiger beetle, which I had never heard of and would not miss if it entered the Valhalla of extinct species.

My cranky reaction was surprising, since I am an environmentalist myself. Nature's web, genetic diversity, and endangered species are sacred to me. I was imagining the speech I would have made if it had ever entered my head to urge Plainfield to adopt a town insect. I would be holier than thou. I would trigger the normal human resistance to missionary zeal. The proposal would be laughed out of the hall.

Fortunately, I didn't make the speech; Nancy Mogielnicki did. Everyone laughed, and then we approved the article. Plainfield now has a town bug.

Mogielnicki's speech was brazen, energetic, and funny. It was a lesson to all sanctimonious environmentalists. She undercut the inevitable ridicule by providing plenty of her own. "What," she asked, "would it *mean* to have a town insect? It can mean whatever we want it to." Given the agility, swiftness, and attractiveness of the insect, she said, we might want to call our athletic teams the Plainfield Tiger Beetles. We could sell tiger beetle T-shirts to raise money for the town Patriotic Committee. Someday, she suggested, we might want to go on and adopt the inverted wedge mussel as the town mollusk.

She made the tiger beetle silly, fun, noble, and an object of town pride all at once. She never mentioned hydroelectric dams. By the time she was done, we all had a kind of warm feeling for the tiger beetle, though none of us, including Mogielnicki, had ever seen one.

The town meeting was just the beginning. A tiger beetle exhibit was put in the school so the kids could become familiar with their town insect. At our July 4 celebration half the town was wearing tiger beetle T-shirts. A gold tiger beetle pendant was auctioned off. And we joined together for the first official singing of our new beetle anthem (to the tune of "The Battle Hymn of the Republic," of course).

The cobblestone tiger beetle's coat is black and bronzy, too.
We deem it far more beautiful than those of emerald hue.
If you think our beetle homely, well, we think the same of you.
Our beetle crawls right on.
Glory, glory, it's the bug we like to see.
Glory, glory it's the bug for you and me.
I'd do anything to find one, even climb up in a tree.
Our beetle marches on.

There are more verses, but you get the idea.

In a more serious mood, another environmentalist called me the other day to complain about how the nation was going down the tubes (we environmentalists often talk like that). He was worried about the younger generation, which, he thinks, is growing up with a materialist ethic, not an environmental one. How can we teach them, he wondered, to treasure and protect the creatures of the earth?

His question got me thinking about how I learned to do that. No one ever preached ecology to me, but my mother took me out to the fields and forests and introduced me to the wildflowers. Those expeditions were fun. I came to feel at home in the wild places. I had friends there, whose names and habits I knew. When the bulldozers came to clear the way for condominiums, I could measure the loss in wild strawberries, spring beauties, and trilliums. My guess is that one needs some such experience of the earth's wondrous variety before one is willing to defend it. You can't treasure something you know nothing about.

In her lighthearted way, Nancy Mogielnicki has given Plainfield such an experience. Now we know the cobblestone tiger beetle's name, what it looks like, and where it lives. That bug is real to us, and we have a personal stake in its welfare. Though some folks think the whole beetle business is pretty corny, I suspect the town would rise up in outrage if anyone tried to mess around with Burnap Island.

The environment is, of course, a serious matter. It is God's creation, the source of all life and wealth. If we stopped to think about it, we would be wonderstruck by the very existence of cobblestone tiger beetles and trilliums, and we would realize the stupidity and arrogance of destroying them. (You see, I can't resist an opportunity for a sermon.) But that doesn't mean the environment must be a heavy trip. Nature can be celebrated while it is being conserved; in fact it may never be conserved until it is celebrated. Let's lighten up.

How do you say "cobblestone tiger beetle" three times very fast without making a mistake?

Bugbugbug.

UNCONVENTIONAL
ECONOMICS

I SPENT considerable time in the Soviet Union and the nations of Eastern Europe before *glasnost* and *perestroika*. Those places always made me sad and angry—sad at the sufferings of my friends there, angry at the stupidity and shabbiness of the economic system and the inhumanity of the ruling powers. Plumbing fixtures all over Russia leak and drip. The air in Budapest in winter clogs the lungs. On the Czech-Austrian border the well-tended, green Austrian fields run right up against the yellowing, sickly Czechoslovakian ones. Stores run out of soap, restaurants run out of beer. People have to live with these small daily indignities, and, of course, worse ones. Worst of all in my experience have been the looks on the faces of my scientific colleagues when they mouthed party lines that they knew and I knew were lies.

I am no friend of Marxism-Leninism. I am also, to the mystification of my right-wing friends, no friend of untrammeled capitalism. I don't see how anyone who cares deeply for the environment or for the well-being of all people can be an uncritical booster of either of these ideologies. From a systems point of view both have obvious flaws: the absence of market feedback in communism, the absence of market control in capitalism, the absence of any consideration of environmental costs in either. Surely we can invent something better or some creative mixture of the two that captures the best of each and rejects the worst.

I've criticized both systems over the years, but there seems little need at this time in history to rehearse the failures of communism. So I have included here only some comments on capitalism. These have been without question the most unpopular columns I have written. Many editors simply didn't publish them. Here a deep paradigm is being threatened. People are viscerally afraid to question the economic ground they stand on.

But we must question it; it is shaky ground, and we can find a firmer foundation if we try. One of the most essential steps toward a sustainable world, I believe, is a rethinking of economics. We have not yet come up with a way of doing our business as though people and the planet mattered. Until we do, people and the planet will suffer, and as they suffer, business will suffer too.

The articles on the long wave included here were written before the stock market crash of 1987. The article on layers of causes was written just after that crash.

Should We Be Glad When the GNP
Goes Up?

WITH THAT special, kindly sparkle in his eyes President Reagan says in his news conference that there are good economic tidings. The GNP figure has been revised upward to a growth rate of 4.5 percent. That is supposed to be cause for rejoicing, for a rising stock market, and for gratitude to the party in power.

No index of national progress—not Olympic medals, not the unemployment rate, not the SAT scores of high-schoolers—is watched more carefully than the GNP. Economists forecast it. Government statisticians revise and polish it. Prime time newscasters regularly alert us the day before a new GNP figure is going to be released.

So quick now, what does GNP stand for? What does it mean? Why should we be glad when it goes up? I wonder how many Americans can answer any of those questions, especially the last one.

GNP stands for gross national product. It means the dollar value of all the final goods and services purchased by the nation's consumers, government, and investors. Here are a few examples to illustrate the many kinds of economic activity the GNP represents and the ridiculousness of counting its every increase as good and its every decrease as bad.

A couple gets divorced and pays a lawyer a hefty fee (GNP up, good). The kids shuttle between his household and hers, requiring duplicate sets of bedroom furniture, toys, and clothes (up, good). She finds cooking for herself too depressing and begins to live on junk food (GNP up, good). He starts spending his spare time fixing up the house instead of hiring someone else to do it (GNP down, bad).

A new lightbulb comes on the market that gives the same light with only half as much electricity; everyone's electric bill goes down (GNP down, bad).

A town decreases its use of salt on the winter roads (down, bad), which causes cars to last two years longer before they rust out and have to be replaced (down, bad). However, more accidents cause an increase in repair bills for cars and people (up, good).

A community floats a $30 million bond for a trash incinerator, which doubles the cost of garbage disposal. New air quality regulations then require more expenditures for scrubbers. The community becomes embroiled in litigation about the disposal of the toxic ash from the plant (up, up, up, good, good, good).

The government decreases highway maintenance (down, bad). It builds more nuclear weapons (up, good). It gives a big raise to Congress (up, good). It eliminates half its paperwork (down, bad).

The GNP is obviously not a measure of progress. It is a measure of monetary flow, effort, expense. Farmer and writer Wendell Berry has called it the "fever chart of our consumption." It is indiscriminate. It lumps together joys and sorrows, triumphs and disasters, profundities and trivialities, everything that costs money and nothing that doesn't.

The GNP measures environmental damage only if we pay to clean it up. It does not register the gardens we grow, the cooking, repairs, and cleaning we do for ourselves. The GNP contains no information about justice. It does not tell us that the number of homeless families has increased and so has the number of families with second homes.

An increase in GNP is good only in the sense that when money is spent, someone gets it, and that someone is usually happy about it. Whether it is good in the larger, societal sense depends on who spent it, who got it, what it bought, and what parts of the transaction were not accounted for.

Economists are well aware of the inadequacy of the GNP as a measure of welfare; they point it out in every textbook. But many economists, like many presidents, forget the caveats and turn into cheerleaders, urging the GNP up, helping to reinforce the national illusion that a bigger economy is a better one.

The problem with that illusion is that it dominates policy. We assume no changes are needed when GNP growth is high; we call on extreme measures when it is low. With gross national product as our indicator, we are in danger of producing gross national product instead of what we really want—health, education, security, a clean environment, jobs with dignity. Surely those goals are more important than just swelling.

When we hear that the GNP has grown, instead of cheering, we should ask exactly what has grown, for whom, at what cost, and at whose expense. Even better, we should work to develop indicators of national progress that reflect more accurately our real values and our real welfare.

A Conservative President Every Fifty Years

NOT MUCH more than a decade ago it was considered normal to raise taxes to balance the government budget. Cuts in military spending were imaginable. There was no real doubt that government should help the poor.

Twenty years ago civil rights, economic equity, and protecting the environment were top national priorities. Rampant militarism and the accumulation of personal wealth were distinctly unfashionable.

Thirty years ago, though our economic output was much smaller than it is now, we were confident enough to talk about ending world poverty and going to the moon.

The tremendous political shift over three decades, from expansive liberal values to cautious conservative ones, has happened not only in the United States but in much of the Western world. Why?

There are many explanations going around. The conservative swing was due to the overwhelming charisma of Ronald Reagan or to people finally coming to their senses and realizing the bankruptcy of the liberal philosophy. Or maybe those who lived through the right-wing fling of the 1920s have died off, and the new generation has to learn again, the hard way, that conservatism is mean and hollow.

There was indeed a swing to the right in the 1920s, very like our present one. Such swings occur with regularity in what appear to be forty-to-sixty-year cycles. U.S. party platforms and British Speeches from the Throne (like our State of the Union messages) reveal marked periods of retrenchment, militarism, great concern with the accumulation of wealth, and decreased social reform. In both Great Britain and the United States conservative periods occurred in the late 1920s, the 1880s, the 1830s, and the 1790s.

For example, a Speech from the Throne in 1830 sounds strangely familiar: "It will be satisfactory to you to learn that His Majesty will be enabled to propose a considerable reduction in the amount of the public expenditure, without impairing the efficiency of our naval or military establishment."

Conversely, there were liberal swings of the political pendulum around 1800, 1850, 1900, and 1960. The society became outgoing and reformist, concerned with the poor, with international cooperation, with solving the problems of the world. In these periods everything was booming and nothing was impossible. Here is a quotation from a progressive Speech from the Throne in 1907: "You will also be invited to consider proposals for the establishment of a Court of Criminal Appeal, for regulating the hours of labour in mines, for the amendment to the patent laws, for improving the law related to the valuation of property in England and Wales, for enabling women to serve on local bodies, and for the better housing of the people."

It isn't only politics that comes in cycles. The great economist Joseph Schumpeter noticed a fifty-year technical cycle. Inventions occur at a steady pace, but the *adoption* of inventions into new industries occurs in bursts. In the 1780s it was mechanized textiles, in the 1840s railroads and steel, in the 1890s electricity, and in the 1950s highways, petrochemicals, aviation, and electronics. The technical upwellings correspond to the liberal phase of the political cycle. The conservative phase comes as factories become old and markets for the latest innovations become saturated.

There is a third cycle, congruent with the political and technical ones. That is the economic cycle called the long wave or Kondratief cycle. It is named for a Russian economist who pointed out in 1925 that market economies have decades-long periods of boom and bust. As far as I know Nicolai Kondratief is the only person to have predicted the Great Depression of the 1930s. He did it by reasoning forward from the depressions in the 1790s, the 1830s, and the 1880s. These depressions, which occurred at roughly the same times in North America and Europe, coincided with the conservative swings of the political cycle; they preceded the innovation upswings of the technical cycle.

Most Western economists did not, and still do not, believe Kondratief. His cycle is not easy to see in economic data because it is not the only thing going on in the economy. There are also four-year business cycles, twenty-year building cycles, and overwhelming long-term growth, all overlapping. Fifty-year long waves are barely visible in GNP records; that measure is dominated by long-term growth. But plots of unemployment rates, consumer price indices, or interest rates reveal cyclic surges, not as regular as clockwork, but clearly visible every forty to sixty years.

Now, as economies are faltering in many parts of the world, there is renewed interest in the long wave. Some economists believe that it is very real and that, despite the misleading measures of the GNP and the stock market, we are currently in the depression phase of the wave. Even MIT economist Lester Thurow, who has scoffed at the idea of long waves, wrote recently, "I am often asked whether the financial panics of the 1920s and the Great Depression of the 1930s could happen again. For twenty years I have answered that what happened then could not happen now. Today I would not so answer."

Maybe Ronald Reagan and Margaret Thatcher were not causes but effects—the conservative choices uneasy people make during times of deep economic uncertainty.

If there are indeed linked economic, political, and technical waves and if the economic wave is headed downward, we had better understand what is going on so we can stop or ameliorate the cycle or at least ride the wave with our heads above water. Kondratief never offered an explanation of why long waves might occur. Now there is such an explanation.

Why There Are Long Waves

THE CHIEF proponents of the long wave theory are a group of computer modelers headed by Jay Forrester at MIT. They have been simulating the workings of the national economy for the past fifteen years. During that time they have been the only theorists I know who have not only explained the strange economics of the 1980s but have predicted them. Here is why, they say, there are long waves.

The economy is made of two basic kinds of industries. One makes *consumer goods*, such as refrigerators, shoes, and cars. The other makes *capital goods*, such as steel, machines, and buildings, which the consumer goods industries use to make refrigerators, shoes, and cars. The capital industry supplies the consumer goods industry, which supplies you and me.

In the mid-1940s, when the long wave began its most recent climb, inventories of refrigerators and cars were low because of the 1930s depression and World War II. Consumers were eager to buy. Refrigerator and automobile factories hired more people, and because more people had jobs, even more consumer goods were demanded. The consumer goods industries needed to expand, so they ordered machines and factories from the capital goods industries.

Before long the capital sector was operating at full capacity. Orders for steel, machines, and factories were still increasing. To fill them, the capital sector first had to make steel, machines, and factories for itself.

While it did that, unfilled consumer orders piled up. The pileup caused the capital sector, trying to gauge how many factories it would need, to raise its expansion plans even higher. Therefore, it created more orders for its own steel and machines. The economy went into the boom of the 1950s and 1960s.

The capital sector cannot distinguish between orders that signify a permanent increase in the scale of the economy and orders that come from a temporary need to gear up production. It assumes orders will keep coming and builds accordingly. Each firm tries to increase its market share and so expands optimistically. Unemployment is falling and wages rising, so the more expensive labor is replaced by capital, which further increases capital-sector orders. Real interest rates are low, so it's easy to borrow for expansion. Above all, there is a general mood of confidence that encourages investment.

The result is overbuilding, especially in the capital sector. Many more steel and machine-tool factories and power plants are built than the economy actually needs.

As the long wave approaches its crest, unemployment is as low as it can get; consumers are buying all the refrigerators and cars they want; inventories become overstocked. Factories that have been in the construction pipeline come into production, though now they are not needed. The economic rush slows down, competition gets tougher. At this point the government usually comes up with some scheme (such as tax cuts and defense spending) to push investment further, trying to keep the boom going. No one really wants to believe in an end to the boom. The overbuilding goes on a while longer.

Finally, the economy's overextension becomes obvious. Inventories pile up, factories are shut down, the long downward slide begins. It will last until enough capital plant is abandoned or torn down so that productive capacity is once again low compared with demand. That can take ten to fifteen years.

Just as the long wave overshoots on the upturn, it undershoots on the downturn. The mood for investment is bad, and even when it improves, new plants cannot be built immediately. Eventually, though, there are clearly too few factories to satisfy demand. That shortage sets up the conditions for the next upturn.

It may not seem possible to you that investors could so badly over- and then underestimate the needs of the economy. They do so because they have no overview, no way of measuring total demand, of knowing what other investors are doing, or of guessing correctly their own market share. That point was driven home to me by a game the MIT group has devised, in which I played the role of a capital-sector investor, trying to build my own plants to meet the incoming orders of my customers and my own expansion. Though I knew the theory of the long wave, first I overshot. Then I undershot. In the process I generated, in simulated time, a fifty-year cycle.

Welcome to the Depression of the 1980s and 1990s

So ARE we entering a depression like the one sixty years ago?

Not necessarily, says Jay Forrester of MIT. The 1930s depression began with a stock market crash; it led to 25 percent unemployment; it coincided with a major drought in the Midwest. The present downturn may not be that sudden or severe. It could be more like the downturn of the late 1800s, which consisted of a series of severe recessions, interspersed with short recoveries. For fifteen years the economy was sluggish, but history does not call it a Great Depression.

The course of the present downturn, Forrester and his group say, depends partly on the mechanics of the long wave itself, and partly on us. The long wave mechanics are the easiest to predict.

1. Long wave downturns are set up by twenty to thirty years of systematic overinvestment. There are now more factories than needed, especially in the heavy industries—metals, construction, machine tools, cement, engines, boilers, and such. That means tremendous competition in these industries worldwide, as we are already seeing. Since U.S. plants are among the oldest and least efficient, we will get more than our share of the shutdowns—we already have.

2. The real interest rate (the nominal rate minus the rate of inflation) rises at the beginning of a long wave downturn for two reasons. Debt, especially government debt, increases, raising the demand for credit. At the same time inflation slows down, and some prices actually deflate, as happened in the 1980s with oil, metals, and farmland. Though nominal interest rates drop, prices drop faster, keeping real interest rates much higher than they were in the inflationary 1970s. They will not go down quickly.

3. Little borrowing is invested in new productive capital. It is credit-card debt for consumers, government debt for military expenditures, junk bond debt for companies to swallow one another up, and foreign debt to pay for imports. The debt is not backed up by real assets but by the vague hope of future economic growth. During those ten to fifteen years, that growth will not happen. Much of the debt will be defaulted. Much Third World debt is being defaulted through the polite process called rescheduling. Junk bond debts and many mortgage debts are also being defaulted, leading to an unprecedented rate of bank failures.

4. Previous long wave downturns occurred primarily in the market economies of Europe and North America. Now the world economy is tightly interlinked. Much heavy industry, which is hardest hit by a long wave, is in Japan, Korea, Brazil, and the oil fields of Mexico, Venezuela, Nigeria, and the Middle East. This depression is already a more worldwide phenomenon than the last one.

5. The future, as seen by the MIT analysts, is not all bleak. Long wave downturns are times of renewal. Increased competition weeds out obsolescence and improves efficiency. It seems to be causing some corporations to try new types of worker participation in quality control, innovation, and management. The downturn also clears the way for the next rise, which will probably begin in the late 1990s.

6. The long wave upturn brings with it widespread adoption of new technical innovations. At the peak, when the economy is capital-rich, there is little room for major capital shifts into new technologies. But at the trough, when capital is short and new facilities are in sudden demand, whole new industries get their start. The IBMs and Xeroxes of the future are forming now around industries like solar energy, bioengineering, satellite communications, waste management, and probably ideas that nobody takes seriously yet.

What is least predictable about a long wave downturn is the social reaction to these economic events. The beginnings of downturns have been characterized throughout history by conservative politics. The wild card is what happens after that, when the economic crisis becomes unmistakable. The policies that seemed to work for thirty years become discredited. Every comfortable, safe belief is thrown into question. Even people who are not directly hurt by the downturn lose confidence.

Uncertain, insecure people can be attracted to any loud voice that sounds sufficiently reassuring. In the 1930s the loud voices were those of Roosevelt, Hitler, and Mussolini. In the 1990s who knows whose they will be. There are plenty of contenders, some of them downright scary.

If the MIT group is right that we are entering a phase of economic depression and political confusion, the best hope for sanity is a clear understanding of what is happening and why. The most important part of the long wave theory is that downturns are no one's fault. They result from rational business decisions made over decades of expansion, from which most people benefit. We do not have the option at this point in the cycle of stopping the downturn. We do have the option of mitigating it by distributing its burdens equitably and by choosing leaders who bring out the best in us, not the worst.

Three Types of Truth About the Long Wave

A TYPE 1 truth is one that remains true no matter what you think or say about it. For example, Halley's Comet reappears every seventy-six years.

A Type 2 truth is more likely to be true the more you say it. "Yes, I can run a marathon," for instance, or "Every child must have a cabbage patch doll."

A Type 3 truth is *less* likely to be true the more you say it. "I'm about to sneeze," or "There is going to be a surprise attack on Tehran next week."

There are few Type 1 truths in economics. Most economic happenings, from stock market gyrations to projected federal deficits, are determined by subtle mixtures of Type 2 and Type 3 truths, and it isn't always easy to tell them apart.

Therefore, one thinks hard before asserting that the world is now in a major depression, a downturn of the Kondratief cycle or long wave, which could go on for another ten or fifteen years. If that is a Type 2 truth, it is better left unsaid.

I am convinced, however, that the long wave is a Type 1 truth, based on the physical and informational structure of our economy. Talking about the long wave will not make it appear or disappear. But there are Type 2 and 3 truths in our social reaction to long wave swings. There I believe that talking about, understanding, and expecting the downturn can very much reduce its negative effects.

I got a letter a while ago from a friend, also a believer in the long wave theory, describing one possible outcome if the downturn comes as a surprise. This scenario need not happen, but it could, if we are not prepared to prevent it.

He said, "I had lunch Monday with a junk bond trader at [a major Wall Street firm]. Our conversation shocked me."

> We talked about the prospect of a financial panic caused by massive defaults on debt. He said there is no way the foreign, farm, mortgage, etc., debts can be repaid. He thinks there will be a panic, followed by some kind of "restructuring."
>
> He was sure that debts will be forgiven. In his scenario the government will simply knock a zero off the balance sheet of all banks, issue a new currency, and in effect default on its own debt. Those who were net lenders would lose 90 percent of their wealth, while those who were net debtors would default on 90 percent of their debt. Your average guy, he said, with a $200,000 mortgage and a nice pension will end up with less pension, but only a $20,000 mortgage.
>
> I asked him about those who are prudent, who don't load up their Visa cards, who live within their means. He admitted that some people would lose jobs and savings, but this was all part of the adjustment process. The glib way he discussed this "adjustment" was appalling. His concern was how he could take advantage of it. His strategy was to leverage himself to the hilt, live it up, maybe buy

some gold as a hedge, and count on the fact that his debts will be forgiven along with everyone else's.

He may be right that political pressure will favor forgiving the debts. But it will be such an injustice if the folks who saved instead of borrowing suffer for the sins of the avaricious. The way to prevent that, of course, is to limit the dollar amount the government would forgive each individual and work out repayment schedules for the remaining debt. That will only happen *if it is decided ahead of time, before the panic and the pressures begin.*

That junk bond trader is playing with a dangerous Type 2 truth. Financial panics happen because people think they will, not because they are inevitable. The current level of debt is indeed too high; some of it is already being defaulted. But a financial panic is not inherent in a long wave downturn, unless we panic.

The long wave downturn is nothing but a correction for excess capital in the economy, especially in the capital-producing heavy industries. Our real wealth—land, resources, people, machines, and know-how—is all still here. The economy is undergoing a structural shift. Some businesses and some workers will be hurt in that shift, if there is no social commitment to help them. But in this rich country no one need experience serious scarcity if we manage properly.

Our main problem will be disappointed expectations. Our aspirations have been built during thirty years of long wave expansion. Many of us have lived our entire adult lives with real incomes rising each year, houses appreciating in value, and debts being paid off by a combination of inflation and economic growth. It's not going to be like that for a while. Once we accept this fact, we'll discover that our survival is not at stake, only our expectations.

The Type 2 and 3 truths about the long wave downturn—what our minds make of the situation—can mitigate the hurt or magnify it immensely. It will be magnified if we permit breakdowns in important social contracts, in the banking, welfare, or trade systems. It will be mitigated if:

• As many debts as possible are honored, though the process may take longer and the interest paid may be lower than anyone now expects. Above all, people must know and trust that their basic savings are protected.

- There is active assistance for those whose jobs are tied to overbuilt economic sectors, so they can shift along with the economy, with their productivity and self-esteem intact.
- We refrain from blaming anyone for our troubles—Russians, Arabs, Mexicans, Republicans, Democrats, junk bond traders, or any other handy scapegoats. We should strengthen, not weaken, the bonds of domestic and international cooperation.
- We carry on a public discussion of the history and theory of the long wave, so nonexperts understand it and so we can work out policies that can, in the long run, reduce the wave's amplitude.

The Stock Market Crash—Layers of Causes

B Y NOW we've heard hundreds of theories about the cause of the 1987 stock market crash. It was the greed of the yuppies that did it, or the German mark, or the August trade balance. It was program trading. It was the government deficit.

We love single, simple causes, though we know that in this complex world causation usually comes in *layers* of increasing profundity. On the surface some person or thing, some flaw or rumor, may have triggered the crash. like the tiny disturbance that causes a bubble to burst. Knowing exactly which pinprick did it gives us no help in preventing another crash—we will never rid the economic world of pinpricks. We have to look deeper and ask why the market was distended and vulnerable in the first place. What blew up the bubble?

Causes of market inflation are not hard to find: tax cuts for the rich that release cash for speculation, financial inventions that allow people to speculate with money they don't have and then to speculate on the speculation, mindless cheerleading from those in authority, who fabricate reasons for the market to increase when there are none.

Those causes are satisfying because they allow us to be righteous as well as partly right, but they don't explain why these inflating forces flourished just now and not in previous decades, when greed and short-sightedness were presumably just as inherent in human nature. To find out why *now*, we have to look even deeper.

Which brings us to what I believe is the most fundamental cause of the 1987 crash.

Economic ups and downs, panics, booms, depressions, have happened regularly for at least 200 years. The worst downturns came in the 1830s, 1880s, 1930s, and 1980s, no matter whether there was a Securities and Exchange Commission or a Federal Deposit Insurance Corporation, no matter whether computers or people did the trading, no matter what party was in power, no matter whether there was a federal deficit. They have happened in every part of the world where there was an industrialized market economy.

We are told, over and over, that the free market is a sort of natural wonder that guides the economy without need for government interference. But in fact the market system is chronically, inherently unstable. All market economies oscillate, with four-to-seven year business cycles, with longer cycles of construction and commodity production, and with fifty-or-so-year long waves that bring, among other things, financial panics.

The oscillations are inevitable because mutual adjustments of supply and demand are slow. It takes time for producers to respond to shortages or surpluses by adjusting prices, and time for consumers to respond to price changes by buying more or less. It takes even longer for producers to gear up or down and for those responses to percolate through the system to suppliers and then to suppliers of suppliers. During the adjustment time—which can be years or even decades—shortages or surpluses go on getting worse until the corrections begin to take hold.

A huge economy cannot bring itself to a prompt supply-demand balance like the smooth graphs in Economics 101 textbooks. Production and consumption yaw back and forth, seeking equilibrium, overshooting, correcting, undershooting, in cycles that impact everything from interest rates to opinions about the president to expectations in the stock market.

By the late 1970s the world's linked market economies had built themselves up to a point of tremendous surplus. Cars, machines, grain, milk, steel, electricity, and oil were being produced faster than they could be sold. Mills, mines, and oil wells were producing at only 50 to 70 percent of their capacity. There was no point in building more of them. Conservative governments cut taxes and created incentives so investors would shore up the faltering economy. But there were few real investments worth making. So a lot of money went into bubble blowing.

When the long wave is going up (as in the 1950s and 60s), fancy speculative devices are unnecessary. Savings can be invested in real growth. Not so when the wave turns down. That is when people dream up sterile mergers and acquisitions, investment trusts, junk bonds, stock futures, and index arbitrage—anything that can keep apparent wealth swelling even though real wealth is stagnant.

Eventually dreams and reality have to be reconciled, and that means some kind of crash. It happened in the 1920s and the 1980s, and it will happen again sometime around 2030, unless we come to understand the economy at its deepest causal layer.

That means putting aside both our eagerness to blame and our knee-jerk ideologies.

If our economic system itself is the deepest cause of the crash, we have no one to blame. Most of us participated joyfully in the ride up. We can minimize the pain only by being compassionate with one another during the slump down.

And if we are serious about understanding our economic system, stabilizing it, and bringing it under some kind of control, we'll have to start by admitting its imperfections. Doing so does not necessarily throw us from the roller coaster of untrammeled free enterprise straight into the gray, unproductive prison of central planning. There's a lot of space in between, in a region that could be called informed, regulated free enterprise. That means no easy labeling of either business or government as all bad or all good. It means finding the right balance between the two.

We can find that balance if we wake up, think, question, reject simple causal theories, and seek to understand at the deepest levels how our economy works.

The Grass Doesn't Pay the Clouds for the Rain

I N THE current vocabulary of condemnation there are few words as final and conclusive as the world 'uneconomic'," wrote E. F. Schumacher in *Small Is Beautiful* (1973). "Call a thing immoral or ugly, soul-destroying or a degradation of man, a peril to the peace of the world or to the well-being of future generations; as long as you have not shown it to be 'uneconomic' you have not really questioned its right to exist, grow, and prosper."

If we human beings are ever going to live in happiness and harmony with one another and with the natural world, we will have to rethink our economics—starting with downgrading the importance of economics in our thinking.

Buckminster Fuller: "The world doesn't run on money. The grass doesn't pay the clouds for the rain."

Mark Sagoff, in *The Economy of the Earth* (1988):

> The things we cherish, admire, or respect are not always the things we are willing to pay for. Indeed, they may be cheapened by being associated with money. It is fair to say that the worth of the things we love is better measured by our *unwillingness* to pay for them. . . . Love is not worthless. We would make all kinds of sacrifices for it. Yet a market in love—or in anything we consider sacred—is totally inappropriate. These things have a *dignity* rather than a *price*.

The way we price things has nothing to do with their value. The natural world in particular, the land and its resources, bears a relationship to economic pricing that is actually perverse. The more valuable the land in market terms, the more destroyed it is in land terms.

Marilyn Waring, in *Counting for Nothing: What Men Value and What Women Are Worth* (1988):

> I turn . . . to the mountains. If minerals were found there, the hills would still be worthless until a mining operation commenced. And then as cliffs were gouged, as roads were cut and smoke rose, the hills would be of value—the price the minerals would fetch on the world market. No price would be put on the violation of the earth, or the loss of beauty, or the depletion of mineral resources. That is what value means according to economic theory.

A new economics, one that could guide us to a world that is both desirable and sustainable, would not be so mesmerized by numbers, money, prices, bottom lines, *quantity* that it forgets to take into account that crucial but unmeasurable characteristic we call *quality*.

Wendell Berry, in *Home Economics* (1987):

> We may transform trees into boards, and transform boards into chairs, adding value at each transformation. In a good human economy, these transformations would be made by good work, which would be properly valued and the workers properly rewarded. But a good human economy would recognize . . . that it was dealing all along with materials and power that it did not make. It did not make trees, and it did not make the intelligence and talents of the human workers. . . .
>
> The good worker loves the board before it becomes a table, loves the tree before it yields the board, loves the forest before it gives up the tree. The good worker understands that a badly made artifact is both an insult to its user and a danger to its source.

A new economics would not madly pursue the growth of everything at any cost. It would realize that on a finite planet perpetual growth is impossible and that an economics for the long term must encompass the concept of *enough*.

Schumacher, again in *Small Is Beautiful*:

> A way of life that bases itself on . . . permanent, limitless expansionism in a finite environment cannot last long. . . . Its life expectation

is the shorter the more successfully it pursues its expansionist objectives. . . . The cultivation and expansion of needs is the antithesis of wisdom. It is also the antithesis of freedom and peace. Every increase of needs tends to increase one's dependence on outside forces over which one cannot have control. . . . Only by a reduction of needs can one promote a genuine reduction in those tensions which are the ultimate causes of strife and war.

Herman Daly and John Cobb, in *For the Common Good* (1989):

Economics can rethink its theories from the viewpoint of person-in-community (and in nature) and still include the truth and insight it gained when it thought in individualistic terms. It need not "junk" its axioms. Many of them can continue to function, only with more recognition of their limits. The change will involve . . . correction and expansion, a more empirical and historical attitude, less pretence to be a "science," and the willingness to subordinate the market to (higher) purposes.

Many thoughtful people are trying to work out a new economics, as though people and the environment mattered. Some of the best of them and their works have been quoted here.

LONGING FOR LEADERSHIP

THOUGH "THE Global Citizen" is an opinion column that appears on the most political page of the newspaper, I try not to make my writing overtly political in the Republican-Democrat sense. Neither party comes close to the kind of platform I'd like to see. I find little evidence in modern politics either of the government for the people our founding fathers envisioned or of the government for the environment that will be necessary for a sustainable world. With regard to the political choices we are typically offered in our elections, my usual response is: a pox on all their houses!

Yet anyone who tries to see the world system whole is bound not only to notice politics but to see in it a fulcrum of obvious power. Not power in the traditional sense—the ability to spend billions of dollars and mobilize mighty armed forces—but power in the systems sense, the power of information/goal setting/leadership. I spend a lot of time pondering that power.

How is it that a nation of 280 million stalwart Russians can be changed completely when just one man, a Stalin or a Gorbachev, is changed at the top?

How can the replacement of just one president for another suddenly turn a nation of 240 million opinionated, free-thinking Americans sharply to the right—or the left?

What is good leadership, anyway?

Why does it seem so absent these days?

Is the human longing for leadership a legitimate need or a refusal of our individual responsibility to find the leaders within ourselves?

From a systems point of view leadership is crucial because the most effective way you can intervene in a system is to shift its goals. You don't need to fire everyone, or replace all the machinery, or spend more money, or even make new laws—if you can just change the goals of the feedback loops. Then all the old people, machinery, money, and laws will start serving new functions, falling into new configurations, behaving in new ways, and producing new results.

Jay Forrester once remarked that no matter what the U.S. income tax laws are and no matter what the welfare expenditures, income distribution remains about the same, just at the edge of what is commonly seen as tolerable inequity. What a leader can do—as Reagan so aptly demonstrated—is work on the socially shared mindset to shift the tolerable inequity. The tax laws and the welfare programs then follow. The same is true for the shared goals of environmental quality, of peace, and of justice. A single persuasive leader working directly on goals and values can shift the functioning of a massive system.

So can a leader who opens up or closes down, speeds up or slows down, distorts or clarifies information flows. That has been the lesson of *glasnost* in the U.S.S.R.

So on the few occasions when I do write directly about politics, I keep coming back to the topic of leadership, sometimes trying to invoke better leadership from the politicians, sometimes trying to invoke it from the public. After all, in a democracy leadership is, or should be, a feedback process, starting from the people, who set goals in their selection of leaders, who by their every speech and action redirect or affirm goals for the people, who then elect new leaders—and so forth.

What Makes a Great Leader?

WAS RONALD Reagan a great leader? I asked that question of a class of Dartmouth freshmen during a seminar on leadership. We had read biographies of Thomas Jefferson, Abraham Lincoln, Mahatma Gandhi, Eleanor and Franklin Roosevelt, and Martin Luther King, Jr. Each student had also studied one leader of his or her own choice—the choices ranged from Peter the Great to Lawrence of Arabia to Lee Iacocca.

You'd think we would have been able to evaluate a recent president whose works we had ourselves witnessed. But of course my question started a fight. Some thought Reagan was one of the greatest leaders of this century. Others thought he was a disaster.

Our problem was the word "leader." We were using it to mean many different things. We needed to take apart all the concepts buried within that word. Like the Eskimos, so expert in snow that they use dozens of words to distinguish different kinds, we needed words to express different kinds of leadership.

To most people "leader" means someone who has power—a head of a state, an army, a major corporation. We decided to call such a person a "ruler." Rulers know how to gain and use power to force things to happen. Peter the Great was a ruler, as was Adolf Hitler, as was Lincoln, who unabashedly revoked the rights of free press and habeas corpus in his determination to win the Civil War. Lincoln was one of the greatest power wielders of all. Reagan rated high with us on the scale of rulership.

We decided to save the word "leader" for a person other people follow not by force but voluntarily. He or she has charm, charisma, fascination, credibility. By that definition Gandhi, King, Hitler, and Reagan all stand out as leaders.

We needed the word "manager" to designate people who know how to organize things, keep the machinery humming, pay attention to details, delegate responsibility. Iacocca struck us as a great manager. So did Eleanor Roosevelt. But not Ronald Reagan. He tripled the national debt; he was unconcerned with the mechanics of government; many of the people he hired were lacking in competence and/or integrity, and either he didn't care or he didn't notice.

The three words, "ruler," "leader," and "manager," cover only the potential of leadership: whether the instruments of power are in hand, whether anyone is following, whether the bureaucracy works. We also needed words to express where the leader is *leading to*—around in circles, or to the promised land, or over a cliff.

A "visionary" is someone who does more than perpetuate the status quo, someone who can articulate a goal so concretely that people create it as a reality. John Kennedy envisioned landing on the moon and rallied the nation to do so. King moved the nation with his vision: "My four little children will one day not be judged by the color of their skin but by the content of their character." Reagan was definitely a visionary, we decided, for a materially prosperous, internationally dominant, old-fashioned moral America.

Hitler was also a visionary, but his vision was deluded and evil. To express a leader's sense of reality, we chose the word "savant." A savant is learned and well informed; he or she does not live in a simplified dream world but comprehends the complexity and variety of the real world. And we chose the word "guru" for a moral leader, one who stands for good and whose presence inspires good in others.

Jefferson was a savant, a well-traveled man who understood the seething ideas and politics of his era; he mastered science, music, agriculture, architecture, and democracy. Lincoln was a self-educated savant, one who had the intellectual confidence to include differing factions in his own cabinet to be sure he would hear all ideas. Reagan failed utterly as a savant. He had little grasp of or respect for facts; he listened to only one side; he saw the world as far more simple than it really is.

Gandhi, of course, was the great guru of this century. Jefferson was a moral leader, as were Eleanor Roosevelt and Martin Luther King. Reagan we rated as pious, not moral. He used the language of morality, but he spoke for special subsets of people rather than for all humanity. Around him factionalism, violence, and greed flourished, not compassion, generosity, or peace.

Once we separated the idea of leadership into these dimensions, the class ended up in agreement, not only about Reagan but also about the other leaders we had studied. The table below summarizes their ratings: ★ means strong; + means adequate; − means deficient. If you agree with these rankings, you can see why Eleanor and Franklin Roosevelt made a good team and what a set of opposites our nation chose when we switched from Jimmy Carter to Ronald Reagan. You may also be shocked, as we were, to see which other leader Reagan most resembles.

	Ruler	Leader	Manager	Visionary	Savant	Guru
Jefferson	+	+	+	★	★	★
Lincoln	★	★	★	+	★	+
Gandhi	−	★	−	★	★	★
E. Roosevelt	−	−	★	★	+	★
F. Roosevelt	★	★	+	+	+	−
Hitler	★	★	+	★	−	−
King	+	★	−	★	★	★
Carter	−	−	+	−	★	+
Reagan	★	★	−	★	−	−

If you disagree with these rankings or if you see more words that should be added to the list, then we can begin a discussion that we should be having nationwide. Citizens of a democracy should be as expert about leadership as Eskimos are about snow, however many words it takes.

Silly Season in New Hampshire

U P HERE in New Hampshire what we call "silly season" starts a full year before our first-in-the-nation presidential primary. Hopeful candidates pop up everywhere shaking hands. We're likely to run into senators, governors, even vice presidents at the shopping center, at the town hall, at the local diner. They parade around our rock-bound little state to test out what sort of person we'll vote for in the next presidential election.

We New Hampshirites are used to this game. Some of us apparently relish it since we fight tenaciously to maintain our god-given right to the first primary in the land. But, frankly, I never look forward to primary season. The hoopla is fun and profitable to the state, but I'm tired of it. I wish we could have a discussion about the best possible president rather than a circus to see who can make jokes, avoid flubs, and keep blow-dried hair in place while wearing funny hats.

I'd like, just once in my life, to have a chance to vote for someone I think would be a great president. But the political parties tramping around here asking for my vote aren't likely to give me that opportunity because to my way of thinking a great president is not a creature of a political party or a fabrication of an election campaign.

I'm looking for someone who is willing to speak to me without being coached by public relations experts; someone who is presented as a human being, not marketed like a new flavor of Coke; someone who can admit to making a mistake, whose shoes are not perfectly polished, who can get mad, who thinks out loud, who can answer a question by saying, "I don't know. I'll do my best to find out."

I'd like a candidate who is not pledged to promote just one way of looking at things, not just the Republican or Democratic way or even my way. I'd like someone who can be president of us all, who listens not just to the right or the left, but who realizes that there's some truth and a lot of exaggeration in every point of view and who knows how to search out the truth.

It would help if this candidate had a stand, a moral, not ideological, stand, one that he or she had come to through experience and reflection, a stand so thoroughly integrated with the candidate's identity that he or she could never be false to it no matter what the pressures—the sort of stand for freedom that Jefferson had, the stand for union of a Lincoln, the stand for equity of a Martin Luther King.

I'd like to vote for someone who wants to win in order to serve the people and the nation, not one who wants to win in order to win.

I'd like someone who knows not just about politics and factions, but about the world, other peoples and cultures, the thoughts and dreams of the 94 percent of humanity who happen not to be born in the United States. I'd like my candidate to know about the rest of the world not just from books, not from advisers, but from having been there.

I want a president who can see beyond the statistics, who can identify with housewives and farmers, steelworkers and small businessmen, the unemployed and the poor, as real people, not as voting blocs.

The candidate I'd vote for would treasure the environment and the resources of our country—the soils and waters and air, and also the human beings, and especially the children—and would realize that in them, not in weapons and threats, is our national security.

Most important, I'd vote for a person who not only speaks the rhetoric of peace, but who deeply understands what peace means; a person who enters negotiations not for show, but to come to agreement; a person who defends the interests, security, and pride of this nation, but realizes that no international order can persist that does not serve the interests, security, and pride of all nations.

As I write down this list of ideals, which I keep in my heart but never speak about in public, all the normal denials are coming up. There is no person with all these qualities. If there were such a person, he or she would not be chosen by our nominating process. And if by chance someone like that were nominated for president, he or she would not be elected.

If all those knee-jerk negatives are true, I might as well go into hibernation until "silly season" is over and then cast a lukewarm vote for one of the public relations creations the parties serve up to me.

But if there's even a small possibility that the 240 million souls in this awesomely powerful land could find and elect a great president, then what? Then, I guess, the thing to do is pitch in, reject the shallow posturing, ask serious questions, and get my friends and neighbors to join me in demanding that the parties, the press, and the candidates treat the election process with the dignity it deserves.

We Loved Ronald Reagan So Much

THERE IS a chilling story called "We Love Glenda So Much" by the South American writer Julio Cortazar. Glenda is a movie star. Everyone loves her. Her fans know her films by heart; they study her every gesture and word.

Of course Glenda has bad moments. In some frames she looks less than glamorous, especially as she grows older. You can find fault with some of her acting here and there. An occasional gesture falls flat.

But film can be edited. Every image can be altered if you're willing to take the time to do it. Glenda's fans, because they love her so much, take the time. They acquire master copies of her films and edit them just a little, just to show their idol at her best.

They set up a worldwide network to replace all copies of her films with the revised versions. It keeps them busy, especially since the films are reedited now and then to make them even better.

More and more people come to love Glenda.

The real Glenda, however, becomes a problem. When she makes personal appearances, she isn't quite the Glenda everyone loves so much. She is looking older. Sometimes she's a little awkward, or ill-tempered, or absent-minded. Her fans persuade her to stay home. Eventually, they convince her not to make any more films.

After a long, lovely time, during which everyone enjoys the wonderful Glenda on the screen, the real Glenda gets bored. She comes out of retirement and schedules a new film and a series of public appearances.

The fans are horrified. They hold emergency meetings to decide what to do, though in fact they know. Because they love Glenda so much, they have to keep her unvarnished, unedited self out of the public eye. They have to lock her up forever or get rid of her permanently.

The story ends there.

Ronald Reagan wasn't our first Glenda, but in many ways he was our best. He was tall and dark with movie star shoulders. His voice was presidential, warm, assuring. He was enthusiastic, upbeat, commanding, funny, folksy. Until Iran/Contra he gave every appearance of being in control. He was making America strong and secure, as a president should do, so the rest of us could relax and go about our business.

We loved him, though it was never a secret that a sizable staff constantly edited him to make him into the president of our dreams.

This is also not the first time we have been shaken by revelations that the president we loved was very different from the all-too-human being in the White House. It isn't the first time the atmosphere has shifted so that one day no one dares say anything bad about the president because everyone loves him, and the next day we all admit that we kind of had our doubts all along.

You'd think we'd learn. But something deep and ancient within us yearns for a perfect leader, one who will protect and uplift us, be better than us, show us the way. When we don't have such a leader, we make someone into one. When he turns out not to fit our fantasies, we blame him instead of ourselves.

The European countries wisely retained the old institution of royalty when they took up democracy. They separated the Glenda job, the national symbol, noble, inspiring, and fictitious, from the job of running the government. A patriotic citizen can criticize the prime minister but not the queen. It's a happy combination, satisfying both the craving for royal perfection and the necessity for democratic dissent.

We, however, load both the symbolism and the burden of government onto just one person. Then we confuse the two roles. We take criticism of a policy as an attack on the nation. If the president fails, we entangle the need to correct him with the need to affirm our faith in the system he symbolizes. We fall all over ourselves trying to chastise the president while defending the presidency.

The president, being human, is likely to fail, especially since we have evolved a selection process that emphasizes his symbolic role rather than his ability to govern. We judge him by his ability to act out our national illusions, and if he's accommodating, he does that and puts the government in the hands of people we have never heard of, until they start pleading the Fifth Amendment before congressional committees.

The only alternative I can see, besides finding a royal family somewhere, is to grow beyond our need for a Glenda. Maybe we can accept that we will always be governed by human beings who are complex mixtures of magnificence and fallibility. Maybe we can reject image making at election time and focus on the managerial and moral qualities of the candidates. Maybe we can recognize that we don't need a perfect leader to express the strength of our nation, that a nation's strength is not in its leaders anyway, but in the maturity, productivity, and common sense of its people.

We Don't Need Leadership to Know Right from Wrong

GIVEN THE national confusion on ethical issues from Baby M to the defense of the Persian Gulf, we could use some moral leadership. But if I'm a typical example, I'm afraid we are likely to look for it in the wrong place.

My all-American public school education was not heavy on ethical analysis. In fact, since I took mostly science courses, my moral confidence was systematically eroded. Every day I absorbed strong messages—values have no place in the laboratory; observe what is happening outside you not inside you; feelings have no validity; if you can't see and measure a conscience, then it must not exist.

My training taught me to determine rightness and wrongness from outside, from measurable criteria such as economic profitability, not from the promptings of an invisible, unquantifiable conscience. And my elders provided me with hundreds of examples of how to rationalize glibly just about any act I might want to commit.

Then I was asked to teach a course on environmental ethics. I didn't know how to begin. How could I lead students through the thickets of moral controversy about population growth, nuclear power, and acid rain? And yet what could be more important than to provide them with some ethical grounding?

To prepare for the course, I sat in on philosophy and religion classes. I read books on ethics. I talked to pastors, priests, and gurus. I looked outside myself for moral leadership.

I discovered that that was the wrong place to look. Inside I had known right from wrong all along.

Religions and ethical theories all have lists of moral rules. They boil down to the ones we learned at our mother's knee. Don't hurt people, don't steal, don't lie. Help each other out.

The rules are not the primary authority, say the ethicists. They derive from something we all have within us, a clear sense of rightness, a sense that is given many names. We can get in touch with it whenever we want to. Prayer and meditation are ways—not the only ways—of getting in touch, of listening for moral guidance.

What that guidance says is consistent and simple. You are precious and special. So is everyone else, absolutely everyone. Act accordingly.

Don't do to someone else what you wouldn't want done to you. Don't do what would cause society to fall apart if everyone did it. Try to do what you would want done if you were someone else—a homeless person in New York, a child in Ethiopia, a Nicaraguan peasant, a Polish dockworker.

You don't want your spouse to commit adultery, so don't do it yourself. You don't want to raise a family on a minimum wage, so pay your workers decent incomes. You don't want to live near a hazardous waste dump, so don't create one. If everyone cheats on income tax or insider-trading laws, the government and the stock market couldn't function. So don't cheat.

It's not hard to see what's right. What's hard is to admit how much of what we do is wrong.

Moral confusion is greatest not at the individual level but at the level of nations. Nations involve people too, people who are all as unique and precious as we are. The rules still apply. We don't want Libyan jets sweeping down in the night to bomb Washington; therefore, it was wrong to bomb Tripoli. We don't want Nicaragua to finance hoodlums to shoot our people and destabilize our government; therefore, it is wrong for us to do that to them. Creating weapons that can destroy not only enemy nations but also our own is so irrational that it defies ethical theory. To think ethically you have to be at least sane enough to recognize a wrong when it threatens *you*.

The usual excuse for state-sponsored immorality is that it opposes the evil of others. When the Soviet Union invades Afghanistan, when white Afrikaners oppress blacks, when Qaddafi harbors terrorists, when Chile tortures political dissenters, they are acting immorally. Don't we have an obligation to do something about it?

That's the hardest part of moral theory for me—what to do about the evil of others. I have found Gandhi to be a wise guide here. Oppose evil, he says, with all your might. Use every possible form of resistance and noncooperation. But don't use violence, which sucks you down into evil yourself. You can't fight evil with evil; fight it only with good.

By any ethical theory the basic assumptions of our foreign policy are immoral. Americans are not more worthy than other human beings. Our nation ought not to have its way at the expense of other nations. The existence of evil elsewhere does not justify committing evil ourselves. Not many of our actions in the world are morally defensible.

Moral leadership does not mean someone to tell us what to do. It means someone to help us discover that we already know what to do, someone who can recognize the smokescreens we all throw into ethical discussions to make us feel good about what we know we should feel bad about, someone to keep reminding us that we are special and precious—all of us, every one of us, but none of us more special or precious than anyone else.

Private and Public Addictions

As WE limit the public spaces where smokers can indulge their habit and consider calling out the army to apprehend drug dealers, it might be well to step back and consider the behavior we are trying to control—addiction. Addiction permeates our private and public lives in more ways than we are willing to admit.

The addictions we try to deal with publicly in varying degrees are drug abuse, alcoholism, smoking, and gambling. Addictions to caffeine and sugar are not considered matters of social concern. Then there are addictions we are perversely proud of. On an individual level they include compulsions toward money, sex, work, or violence. On a societal level some examples are government borrowing, oil, pesticides, fertilizers, and militarism.

If that sounds like a strange list, keep in mind what addiction is, a habit you're hooked on though it hurts you, a quick fix that relieves or hides the symptoms of a problem while actually making the problem worse. The fix drives you deeper into a hole while making you feel momentarily better because when you are under the influence, you are out of touch.

The government acts like any common addict in spending money it doesn't have, thereby accumulating debt, thereby incurring interest charges, which it pays by further borrowing. Listen to Congress and the president explain how they plan to reform any time now. They sound just like an alcoholic who swears he can quit whenever he wants to.

Our policy toward oil perfectly fits psychotherapist Anne Wilson Schaef's description of addiction: "When we are functioning out of an addictive process, we will do anything to protect our supply—whatever that supply is."

We send military forces to maintain shipments through the Persian Gulf at a defense cost of $180 a barrel for $20 a barrel oil. We displace people, despoil wilderness, search the bottom of the sea to find the last oil deposits, like a drunk ransacking the house hoping to turn up one more bottle. If we can keep the oil flowing and pretend it's cheap by not counting its environmental or political costs, we can put off facing our real problem, which is how to moderate our energy demands and tap renewable supplies.

If we use enough fertilizer, we can cover over the fact that our farming methods use up the humus, tilth, and natural fertility of the soil. If we can invent new pesticides faster than the pests become resistant to the old ones, we won't have to admit that our monocultures are unmanageable. The smoker doesn't want to know what's happening to his lungs or the alcoholic to his liver, and we prefer not to know how agricultural chemicals are affecting our soils, our waters, our ecosystems, or our own health.

Addiction is ingrained in our society. It is reinforced by every speech or advertisement that tells us to feel good and not look too deeply into reality. Our public discourse encourages us to indulge in substances, sex, or materialism rather than come to grips with either temporary upset or long-term inner emptiness.

The roots of addiction lie in selfishness and self-deception. Once an addiction has started, it takes on a formidable power that cannot be weakened by reason, shame, laws, armies, or ordinary will power. A compulsive overeater can no more "just say no" than can a cocaine addict or a defense contractor going after a fix of public money. They are all capable of any act, no matter how crazy or harmful, that will maintain their habit.

The amazing thing is that addiction can be overcome. This country of addicts has spawned what must be the largest, most effective network of recovery in the world—Alcoholics Anonymous and its many offshoots.

The members of A.A. never speak as a group on public issues, even those such as drugs or smoking that deal directly with addiction. But they will tell you as individuals what has worked in their personal experience. It's a process worth thinking about as we struggle with our nation's addicts and our national addictions.

The first and crucial step is admitting the addiction, the lack of control, the insanity of one's actions.

That step of humility is followed by surrender and honesty—

surrender to one's higher purposes and powers and honesty about one's past and present actions, including restitution wherever possible.

The final step is the maintenance of recovery by helping others to recover.

That may sound like more than an ordinary human can do, much less a crack addict or politician or defense contractor. But all around you are ex-addicts who have done just that. They did it slowly, one day at a time. And they did it with the constant support of a society of fellow addicts, who understood their struggle.

We are all addicts of one kind or another. As they say in A.A., God grant us the serenity to accept the things we cannot change, the courage to change the things we can, and the wisdom to know the difference.

The Map Is Not the Territory, the Flag Is Not the Nation

I FOUGHT FOR three years for that flag. Anyone who burns it must be a traitor."

That was one man-on-the-street's comment in response to the president's call for a constitutional amendment banning flag burning. Most people who were interviewed supported the amendment, many with high emotion.

"The flag is the symbol of our country. No one has the right to burn it."

"People who would even think of doing that are criminals, and they ought to be treated as criminals."

In only one news report did I hear a respondent observe quietly, "It's just a piece of cloth."

Listening to the president's statement and the approving echoes of my fellow citizens, I shuddered. What kind of civic education has our population received? What kind of *basic* education if we can't distinguish between a symbol and reality? Does that veteran really think he fought three years for a *flag* rather than for a nation, one of the few nations in the world in which all citizens can criticize the government, using any symbols they choose?

I suppose every one of us has at some time been furious with our government. The government gives us plenty of reasons to be furious, from illegal wars, to unwarranted infringements of our personal freedoms, to an incomprehensible and unjust tax system. It ripped off taxpayer dollars at U.S. Department of Housing and Urban Development and will be ripping off more to cover the greed of the savings and loans industry. I can understand citizen outrage.

I can't understand, though, why people would express their displeasure by burning a piece of cloth when, in this democratic nation, they have so many more effective actions open to them. I also can't understand why that gesture of flag burning, which harms no one and changes nothing, so enrages other citizens. Sometimes I wish everyone would read a book I use when I'm educating future citizens, S. I. Hayakawa's *Language in Thought and Action* (1939). The symbol is *not* the thing symbolized," thunders Hayakawa. "The map is *not* the territory. The word is *not* the thing."

> Most societies systematically encourage . . . the habitual confusion of symbols with things symbolized. For example, if a Japanese schoolhouse caught fire, it used to be obligatory in the days of emperor-worship to try to rescue the emperor's picture (there was one in every schoolhouse), even at the risk of one's life. . . . The symbols of piety, of civic virtue, or of patriotism are often prized above actual piety, civic virtue, or patriotism. In one way or another, we are all like the student who cheats on his exams in order to make Phi Beta Kappa; it is so much more important to have the symbol than the things it stands for.

That is the best explanation I have heard for how a president could be willing to require people to say the Pledge of Allegiance and to punish people for burning the flag—to use the force of government to strengthen the symbols of freedom while undermining real freedom—and how the people could actually be willing to let him do it.

It's not surprising, I guess, that we muddle symbols and reality. We are told every day that owning a big car makes one an important person, that wearing fashionable clothes makes one more attractive, that having lots of things bought with credit cards makes one well off. We are told that going into debt to buy powerful, expensive, unusable weapons makes us a powerful nation. Some people profit greatly from keeping us confused about symbols and reality.

Says Hayakawa, bluntly:

> We live in an environment shaped and largely created by . . . mass-circulation newspapers and magazines which are given to reflecting, in a shocking number of cases, the weird prejudices and obsessions of their publishers and owners; [and by] radio and television programs . . . almost completely dominated by commercial motives; [and by] public-relations counsels who are simply highly paid craftsmen in the art of manipulating and reshaping our semantic environment in ways favorable to their clients. . . .
>
> Citizens of a modern society need, therefore, . . . to be systematically aware of the powers and limitations of symbols, if they are to guard against being driven into complete bewilderment.

Let us think clearly and avoid bewilderment. The flag is not the government. It is not the nation. The Pledge of Allegiance is not allegiance. People who can't understand those distinctions are in grave danger of ending up with all the symbols of freedom—and with no freedom.

KEEPING GOING WHEN THE GOING GETS TOUGH

My FIRST job when I write my column and when I teach is to awaken people to the basic environmental information and the ever-breaking news. That's the easy part. It's exciting to learn how our planet works, and it's fascinating to follow its story.

My second job is much harder. After awareness comes a state of overwhelm. Students and readers begin to see the problems as monumental and themselves as very small. They see the enormous forces of greed and shortsightedness in the world and the awesome powers of destruction. Their eyes glaze over, not in boredom, but in panic. I have to help them keep a sense of perspective, of possibility, of uplift, of a proper balance between helpless pessimism and sappy optimism.

I have to do it for myself, too.

I try to keep my balance, first, by staying attuned to the good news. There is plenty of it. Because the media are filters that let through mostly bad news, I have to find the good news on my own, not usually among the famous and powerful, but among my own friends and neighbors. It's not hard to do. In my daily experience (as opposed to my experience of the nightly news) the world is much more full of good than of evil.

There are plenty of examples, if you're open to them, of people being wonderful and human affairs working, just as there are examples that lead us to conclude that we are all fatally flawed and doom is just around the corner. The trick is to keep admitting *all* the evidence, all the complexity, much of it contradictory. As some wise person once said, optimism and pessimism are just two different forms of surrender to the same simplicity. The world isn't simple.

A second way I keep myself going is to keep my feet firmly on the ground of present reality but my eyes on a vision of a better world. I can see that vision in some detail. I can see how the energy systems could work, how the farms will look, how materials will be handled, how human needs will be met, how people will care for one another and the earth. Nothing in that vision is impossible. Indeed, I can see it so clearly because I have seen every piece of it actually operating somewhere on earth.

It calms me to stay in touch with my vision because I see that the task of attaining it is quite simple. It would be much simpler, in fact, to bring forth a world in which the interconnected, mutually reinforcing goals of environmental sustainability, economic sufficiency, peace, and justice are reached than it would be to continue on our current stressful, expensive, self-undermining path. Whenever I lose my way, I get myself back to my picture of the world I am working toward. In that vision all the separate topics I write and speak about fall together into harmony—population, energy, materials, land use, a new economics, a new kind of leadership based on the wisdom of the people, and a new (or renewed) morality.

Even if I didn't have a vision, even if there were not evidence of workability right under my nose, I would lean toward an optimistic interpretation of events simply because optimism works better than pessimism. When I'm captivated by hopelessness (and I have my hopeless interludes), I don't do much, and much of what I do is more vindictive than effective. It's only out of a sense of real possibility that I can produce any useful results in the world.

As I was sorting out some of my most upbeat columns for this section, I noticed two things about them that should not have come as a surprise. First, these pieces tended to be some of the most integrative, holistic columns I have written. It seems easier to make a case for workability when I back off to see the big picture, the whole system. Second, these pieces are the most personally satisfying to me. They come closest to expressing who I am and what I'm about—and, I believe, who everyone is and what we're all about.

When people make mistakes, we say, "Well, they're only human." When people act stupidly or selfishly, we say, "That's the human condition," as if membership in the species Homo sapiens were a life sentence to blundering and malevolence. But we also talk about "humane" treatment, and we praise a person's "real humanity," and in so doing we acknowledge the noble essence within ourselves, the godliness in each of us. It is out of that nobility, which every one of us can feel inside and yearns to express, that we can create a sustainable world.

Good Environmental News from All Over

IF YOU pay attention to the news, you can easily get the impression that the waters and air and soils of the earth are deteriorating and that nobody's doing much of anything about it.

You may conclude that protecting the environment is a tough, expensive job. You have to choose—a healthy environment or a healthy economy. You can't have both.

But when I look around, I see a lot of environmental progress. Sure there are plenty of problems—acid rain, hazardous waste, soil erosion. But there are also examples of people treating resources with care, wasting less while producing more, and making money in the process.

For example, here's a collection of environmental good-news stories I ran across in just a few weeks' time.

- The World Bank has gone into the national park business. While financing a large dam on the island of Sulawesi in Indonesia, the bank insisted on protecting 800,000 acres of forest on the watershed above the dam. The trees regulate water runoff and prevent the dam from silting up. Since many endangered species live there, the forest has become a tropical research station, bringing in foreign exchange from visiting scientists.
- In arid Senegal a program of interplanting millet and peanut crops with nitrogen-fixing acacia trees is doubling crop yields and, at the same time, improving soil fertility, reducing erosion, and catching and holding more water.
- Israel has pioneered water-use technologies so efficient that over the decade 1968 to 1978 agricultural output nearly doubled, while water use per acre of irrigated land fell 21 percent.
- Companies are learning that they can make money cleaning up their

271

environmental act. The 3M Corporation in St. Paul, Minnesota, has redesigned its manufacturing processes to eliminate each year 90,000 tons of air pollutants, 10,000 tons of water pollutants, a million gallons of waste water, and 15,000 tons of solid waste. In the process the company saved $200 million. A Goldkist poultry plant came up with procedures that used 32 percent less water and generated 66 percent less waste and saved $2.33 for every dollar spent to institute the changes.

- The average new house in Sweden requires only half as much energy to heat and cool as the average new American house (per square foot, under identical climatic conditions), and it costs less and is more solidly built. Sweden imposes stricter energy standards on construction than the United States but is more flexible in letting builders attain those standards by any structurally sound method.

Some of the best news comes in the field of energy, both in technologies to save it and in renewable, solar-based ways to obtain it.

- The typical top-mount-freezer automatic-defrosting refrigerator on the United States market in 1972 required almost 2,000 kilowatt-hours of electricity per year. The most energy-efficient model now available, with the same features, uses only 800 kilowatt-hours per year. Research prototypes are in operation that use only 250 kilowatt-hours.
- The average 1986-model car in the United States travels twice as far on a gallon of gas as the average new car of 1973.
- Since 1973 the United States has saved the equivalent of 10 million barrels of oil per day through simple conservation measures. Even at low oil prices, that saves us $30 billion every year. The amount of oil saved by conservation since 1973 is five times the amount saved by shifting to coal burning and ten times the amount taken up by nuclear power.
- Since 1980, 13,000 windmills have been installed in California. Together they generate over 1,000 megawatts, the equivalent of one nuclear power plant (and they were built much more quickly than a nuclear power plant could be).
- Low oil prices have not killed off solar technology. Production of solar collectors in the United States is rising at the phenomenal rate of

30 percent per year. Shipments of sun-powered photovoltaic generators tripled between 1982 and 1984.
- Industrial cogeneration (the joint production of electricity and
  steam, which is twice as efficient as electric generation alone) now
  amounts to 15,000 megawatts in the United States, the equivalent of
  fifteen nuclear power plants. At least 16,000 more megawatts of
  cogeneration are currently under construction.

I conclude from stories such as these that the environment can
indeed be improved and that some ingenious people are working hard
to improve it. High productivity in a healthy environment is possible.
But it isn't guaranteed, not until this kind of good news is multiplied
manyfold everywhere on earth.

Problems Are Interconnected—and So Are Solutions

ONE OF the favorite maxims of environmentalists is, "Everything is connected to everything else." That idea is usually
delivered with a heavy charge of negativity: If you do something stupid in one place, it will lead to bigger problems somewhere
else or horrible disasters some time in the future. Your ecological
misdemeanors will come back to haunt you.
  Environmentalists like to collect examples of awful interconnections, some of which you've surely heard:

- Buy a fast-food hamburger and you will create a demand for cheap
  beef, which will be produced in Central American pastures, which
  will be created by chopping down tropical forests, which are the

wintering grounds of the songbirds that glorify your summer and eat your mosquitoes. You will also contribute to poverty in Central America, which will encourage communist movements, which will have to be opposed, which will require constant funding from Congress, which will necessitate frequent speeches from the president, interrupting your favorite prime-time television programs.

- Build a dam on a river and you will stop the annual flood of nutrients that fertilizes the downstream soil; you will reduce the source of food for the fish populations in the estuary, thereby bankrupting the fishing fleet; your reservoir will flood historic monuments, prime topsoils, and endangered species; you will so alter groundwater levels that harmful metals will be leached out of the soil and become lodged in the bones of your children.

- Buy a fast-food hamburger packaged in a foam-plastic container (hamburgers are sources of many ecological disaster stories), and you will release chlorofluorocarbons that will wipe out the ozone layer, thereby allowing the sun's ultraviolet rays to penetrate, thereby frying most terrestrial forms of life. If the hamburger is wrapped not in foam but in paper and cardboard, you will contribute to the destruction of northern forests, which are slowly dying anyway from acid rain from the coal-fired electric plants needed for the ugly roadside signs that attracted you to the hamburger joint in the first place.

The way it is usually told, the message, "Everything is connected to everything else," is not fun to hear. It is intended to cause repentance and reformation. More often, of course, it causes guilt, fear, and an uncontrollable urge to avoid environmentalists.

What we are rarely told is that *solutions* are as interconnected as problems. One *good* environmental action can send out waves of *good* effects as impressive as the chain of disasters that results from environmental evil.

Take energy efficiency, for example. That doesn't mean deprivation of creature comforts; it means insulating houses, driving cars with better mileage, and plugging in appliances that deliver the same service for less electricity.

Energy efficiency is a solution to economic problems—it cuts costs to homes, businesses, and government. We could cut our $430 billion annual energy bill in half just by being as efficient as Japan and West Germany are.

And look at all the other problems efficiency solves. It could allow us to shut down every nuclear power plant in the country, eliminating the need for heroic financing, political hassles, evacuation planning, the disposal of undisposable wastes, and the bureaucracy of the Nuclear Regulatory Commission. It could free us from the need for Persian Gulf oil and from involvement in Middle East wars. It would do wonders for our trade and budget deficits. It would improve the air quality of our cities and go a long way toward solving the problems of acid rain and of global climate change.

Recycling is another favorite environmental solution—a solution to the problem of where to put the garbage. It also reduces groundwater contamination from leaking landfills and air pollution from incinerators. It provides paper, metals, glass, plastics, rubber, oils, and other raw materials for a host of businesses. It slows the depletion of forests and mines and reduces pollution from smelters, oil refineries, paper mills, and plastics factories. It saves a lot of energy and therefore contributes to all the good interconnections listed in the previous paragraph.

Organic farming, practiced successfully by tens of thousands of farmers in this country, can cut a farmer's costs, helping to save the family farm and reducing the need for billions of dollars of farm subsidies. The use of fewer hazardous chemicals improves the health of farmworkers and animals, reduces contamination of groundwater and lakes, restores wildlife populations, and eliminates the need for polluting chemical factories. It saves energy (half the energy used for agriculture goes into the manufacture of fertilizer). Returning organic wastes to the land reduces soil erosion, improves water retention, slows siltation of downstream reservoirs, and reduces urban garbage.

Many of these environmental solutions are considered "uneconomic," but that is because the economics have been figured only for the most short-term and close-in links of the chains. If we calculated the effects on the whole system, we'd see that the wages of environmental sin may be deadly, but the profits of environmental good sense can be enormous.

Everything *is* connected to everything else on this planet. That can be good news as well as bad.

The Senior Citizens of 2089

ANN LANDERS ran a tribute recently to today's senior citizens, who have survived a lifetime of change.

She pointed out that they were here before the pill, penicillin, plastics, and pantyhose; before television, Xerox, and ballpoint pens; before tape recorders, much less CD players; before the forty-hour week, coed dormitories, and computers. The people of our older generation have gone through a lot. They have adjusted to great transformations.

Peggy Streit, editor of *World Development Forum*, was prompted by that tribute to seniors to think about the senior citizens of 100 years from now. What changes will they witness? What adjustments will they have to make? Her imaginings were not cheerful.

They will be the ones who were here before the seas rose to flood coastal plains around the world; before the Panama Canal was silted up; before the headwaters of the Nile became a trickle; before the AIDS epidemic decimated cities on three continents.

They were around when the Amazon forests were something you could visit rather than read about in history books; when venturesome travelers could still see gorillas, elephants, sea turtles, and bald eagles; when only a handful of lakes had gone lifeless; when sunbathing was fashionable and not fatal; when Los Angeles was a megalopolis with swimming pools, irrigated lawns, and free drinking water.

They will remember when the USA and USSR—then called superpowers—still thought the name of the global game was ideology, not ecology.

Those senior citizens, if they are not extinguished in a nuclear winter, may have to be an even hardier bunch than their ancestors of 1989. Unless, that is, nations heed their scientists and begin soon a massive, concerted effort to save their fragile planet.

That vision of a dismal future is a common one these days, and understandably so. An ecologically deranged earth is at the end of the path our current society is hotly pursuing. The reason for describing it, as Streit has done, is to try to whip up some political will, some intention *not* to travel in that direction.

The trouble with such descriptions, though, is that they may stimulate not will but denial. They only point out the way *not* to go. Political will is more readily summoned for a way forward to a desirable future. We need to imagine not only a desolated world 100 years from now, but also one that has achieved sustainability, peace, and justice.

So here's the beginning of another possible tribute to the senior citizens of 2089. I invite you to add to it.

The elders of 2089 were here before the cities were regreened, when the streets were still unsafe, and when people choked on pollution. They themselves helped to reclaim waste spaces, build up topsoil, and plant trees. They witnessed the harnessing of solar energy and the phasing out of polluting machines powered by coal, oil, and nuclear reactions.

They can remember when people actually tried to control pests by spraying poisons over the whole countryside. They helped pioneer the agricultural revolution that produced high yields using elegant natural controls instead of toxic chemicals.

They think back with disbelief on the times of their parents, when people tossed away mountains of paper, metal, and plastic and then went to great expense to wrest more from the soils and rocks of the earth. They remember learning how to recycle; some of them were enriched by investing early in the great materials recycling industry.

They were the first generation to think of the societal implications when they chose the number of children they would have—and they had two at most. They were also the first generation with whom society worked wholeheartedly to assure the sustenance, the health care, and the first-class education of every one of their children.

These elders saw with their own eyes the teams of tree planters reclaiming the deserts and reversing the greenhouse effect. They saw the Amazon forests regenerate and streams everywhere begin to flow clear. They remember the news reports when, one by one, the populations of gorillas, elephants, and bald eagles began to rise again.

They are the last generation to have experienced what it was like when much of the human race was hungry and desperate, when great

areas of the world were poor, oppressed, and angry. They are profoundly grateful that they can now travel anywhere and find sufficiency, stability, warmth, and welcome.

They remember living under the terrifying shadow of nuclear war.
They also remember the historic day when the last nuclear weapons
were dismantled and people everywhere turned out in celebration.
They love to tell stories of that day. They say their lives have been
great, but that was the greatest day of all.

The Man Who Planted Trees and Grew Happiness

I BELONG TO a twenty-nation network of environmentalists and
resource managers, which has taken as its guiding document a
beautiful story about a single Frenchman. Our network consists of
East and West Europeans, Russians, North and Latin Americans,
Asians, and Africans. The story goes to all our hearts. Here it is.

In 1914 Jean Giono was hiking through the barren hills of Provence
in southern France. Charcoal burners had deforested the land, the
streams had dried up, the villages were deserted. Desperately in need
of water, Giono finally found a lone shepherd who gave him a drink
from his well and invited him to spend the night in his cottage.

The shepherd's name was Elzeard Bouffier. He was fifty-five years
old. His wife and son were dead; he lived with his sheep and his dog.

That night Giono watched Bouffier sort out 100 perfect acorns from
a bag. The next day, up in the hills, Giono accompanied Bouffier as he
planted them. Bouffier also tended seedlings of birch and beech that he
had planted in dried-up watercourses. He told Giono that he planted
100 acorns every day. He supposed he had planted 100,000 of them,
20,000 of which had sprouted, one-half of which would live and grow.

"It was his opinion," Giono notes, "that this land was dying for want of trees. He added that, having no very pressing business of his own, he had resolved to remedy this state of affairs. I asked him if the land belonged to him. He answered no. Did he know whose it was? He did not. . . . He was not interested in finding out. He planted his hundred acorns with the greatest care."

Giono went off to fight in World War I and did not return for five years. When he came back, he saw that "a sort of grayish mist covered the mountaintops like a carpet." The oldest oaks were now ten years old, and trees covered an area two miles by five miles. Bouffier had given up his sheep because they threatened the young trees. Now he was keeping bees and still planting 100 acorns a day.

The trees were starting a chain reaction of nature restoring itself. "I saw water flowing in brooks that had been dry since the memory of man. As the water reappeared, so there reappeared willows, rushes, . . . and a certain purpose in being alive."

In 1935 the French Forest Service came by to protect this "natural forest" from charcoal burning. By that time Bouffier was ten miles away at the leading edge of the forest, still planting.

The last time Giono saw Bouffier was in 1945. Bouffier was then eighty-seven, still at work. "Peaceful, regular toil, the vigorous mountain air, frugality, and above all serenity of spirit had endowed this old man with awe-inspiring health. He was one of God's athletes."

The entire region had been transformed. When Giono had first visited in 1914, the few residents "had been savage creatures, hating one another, living by trapping game, little removed, both physically and morally, from the conditions of prehistoric man. All about them nettles were feeding upon the remains of abandoned houses."

By 1945, "Hope had returned. Ruins had been cleared away, dilapidated walls torn down and houses restored. People . . . had settled here, bringing youth, motion, the spirit of adventure. The new houses, freshly plastered, were surrounded by gardens where vegetables and flowers grew in orderly confusion. On the lower slopes of the mountain I saw little fields of barley and of rye."

According to Giono, Elzeard Bouffier died peacefully in 1947 at the hospice in Banon.

A member of my international network from Switzerland showed me this story in German. It was originally written in French and first published in English in *Vogue* in 1954. It has been reprinted many times and in many languages. The German edition I first saw ends with an interview with Giono, in which the interviewer points out that there is no record of Bouffier's death at Banon and asks exactly where Bouffier's forest is.

Giono replies, "If you go to Vergons, Banon, or Le Larne, you will see nothing. Since that time everything has been changed so that silos for atom bombs, shooting ranges, and various oil reservoirs could be placed there. Only a few groves remain. Be content with the story and the spirit of the deed."

I asked the European members of the network to check the authenticity of Elzeard Bouffier. They could find no information. Stewart Brand, reprinting it in the *Whole Earth Catalog* in 1980, asked, "Is the story true? I've heard vociferous denials and confident affirmations." Brand then listed stories of other undoubtedly real tree planters: the Hoedads of California and Wendy Campbell-Purdy, who has planted 130,000 trees in the desert of Algeria. In these cases the same phenomenon followed the tree plantings—streams were restored, soil was improved, wildlife, villages, gardens, and grains returned.

Recently, Chelsea Green Publishing Company issued a new edition of Giono's story with an afterword by Norma L. Goodrich, a scholar of French literature. She states that the story of Elzeard Bouffier is fiction—loving, hopeful, heartfelt fiction.

It is fiction that has inspired thousands of real people to carry out every day their own particular equivalent of planting 100 acorns. "The Man Who Planted Trees" is beloved by my colleagues in twenty countries. It is my own favorite story. Real or unreal, it doesn't matter, it still strengthens me.

"Be content with the story and the spirit of the deed."

Lines in the Mind, Not in the World

T**HE EARTH** was formed whole and continuous in the universe, without lines.

The human mind arose in the universe needing lines, boundaries, distinctions. Here and not there. This and not that. Mine and not yours.

That is sea and this is land, and here is the line between them. See? It's very clear on the map.

But, as the linguists say, the map is not the territory. The line on the map is not to be found at the edge of the sea.

Humans build houses on the land beside the sea, and the sea comes and takes them away.

That is not land, says the sea. It is also not sea. Look at the territory, which God created, not the map, which you created. There is no place where land ends and sea begins.

The places that are not-land, not-sea, are beautiful, functional, fecund. Humans do not treasure them. In fact, they barely see them because those spaces do not fit the lines in the mind. Humans keep busy dredging, filling, building, diking, draining the places between land and sea, trying to make them either one or the other.

Here is the line, the mind says, between Poland and Russia, between France and Germany, between Jordan and Israel. Here is the Iron Curtain between East and West. Here is the line around the United States, separating us from not-us. It's very clear here, on the map.

The cosmonauts and astronauts in space (cosmonauts are theirs, astronauts are ours) look down and see no lines. They are created only by minds. They shift in history as minds change.

On the earth's time-scale, human-invented lines shift very quickly.

The maps of fifty years ago, of 100 years ago, of 1,000 years ago are very different from the maps of today. The planet is 4 billion years old. Human lines are ephemeral, though people kill one another over them.

Even during the fleeting moments of planetary time when the lines between nations are held still, immigrants cross them legally and illegally. Money and goods cross them legally and illegally. Migrating birds cross them, acid rain crosses them, radioactive debris from Chernobyl crosses them. Ideas cross them with the speed of sound and light. Even where Idea Police stand guard, ideas are not stopped by lines. How could they be? The lines are themselves only ideas.

Between me and not-me there is surely a line, a clear distinction, or so it seems. But now that I look, where is that line?

This fresh apple, still cold and crisp from the morning dew, is not-me only until I eat it. When I eat, I eat the soil that nourished the apple. When I drink, the waters of the earth become me. With every breath I take in I draw in not-me and make it me. With every breath out, I exhale me into not-me.

If the air and the waters and the soils are poisoned, I am poisoned. Only if I believe the fiction of the lines more than the truth of the lineless planet will I poison the earth, which is myself.

Between you and me, now there is certainly a line. No other line feels more certain than that one. Sometimes it seems not a line but a canyon, a yawning empty space across which I cannot reach.

Yet you keep appearing in my awareness. Even when you are far away, something of you surfaces constantly in my wandering thoughts. When you are nearby, I feel your presence, I sense your mood. Even when I try not to. Especially when I try not to.

If you are on the other side of the planet, if I don't know your name, if you speak a language I don't understand, even then, when I see a picture of your face, full of joy, I feel your joy. When your face shows suffering, I feel that too. Even when I try not to. Especially then.

I have to work hard not to pay attention to you. When I succeed, when I close my mind to you with walls of indifference, then the presence of those walls, which constrain my own aliveness, are reminders of the you to whom I would rather not pay attention.

When I do pay attention, very close attention, when I open myself fully to your humanity, your complexity, your reality, then I find, always, under every other feeling and judgment and emotion, that I love you.

Even between you and me, even there, the lines are only of our own making.

For Further Reading

All authors stand on great shoulders, especially in the arena of sustainable development, where there has been much superb writing and thinking. The books that have most informed my understanding and my columns are listed here. They supply two types of information: basic conceptual foundations and ongoing statistical updating.

To me the works listed here are more than recommended reading. They are more than what the informed person needs to comprehend to be effective in the world. I think they are treasures. I offer them to you with joy.

Conceptual Foundation

Wendell Berry, *The Unsettling of America*, San Francisco, Sierra Club Books, 1977.

Wendell Berry, *Home Economics*, San Francisco, North Point Press, 1987.

Rachel Carson, *Silent Spring*, original edition Boston, Houghton Mifflin Company, 1962.

Herman Daly, *Toward a Steady-State Economy*, San Francisco, W. H. Freeman and Company, 1973.

Herman Daly, *Steady-State Economics: Second Edition with New Essays*, Washington, DC, Island Press, 1991.

Herman Daly and John Cobb, *For the Common Good*, Boston, Beacon Press, 1989.

J. W. Forrester, *Principles of Systems*, original edition, Wright-Allen Press, 1968, now distributed by Productivity Press, Cambridge, MA.

J. W. Forrester, *Urban Dynamics*, original edition, M.I.T. Press, 1969, now distributed by Productivity Press, Cambridge, MA.

Jean Giono, *The Man Who Planted Trees*, originally published in *Vogue*, 1954, reprinted by Chelsea Green Publishing Company, Chelsea, VT, 1985.

Paul Gruchow, *The Necessity of Empty Places*, New York, St. Martin's Press, 1988.

Garrett Hardin, "The Tragedy of the Commons," *Science*, Vol. 162, 13 December, 1968, pp. 1243–1248.

S. I. Hayakawa, *Language in Thought and Action*, Harcourt Brace Jovanovich, Inc., 1939.

Thomas S. Kuhn, *The Structure of Scientific Revolutions*, Chicago, University of Chicago Press, 1962.

Aldo Leopold, *A Sand County Almanac*, Oxford University Press, 1949.

E. F. Schumacher, *Small Is Beautiful*, New York, Harper & Row, 1973.

Statistical Updating
These are all available in annual or biannual editions.

Lester Brown et al., *State of the World*, New York, W. W. Norton.

Population Reference Bureau, "World Population Data Sheet," available from PRB, 777 Fourteenth Street, NW, Suite 800, Washington, DC 20005.

Ruth Leger Sivard, *World Military and Social Expenditures*, available from World Priorities, Box 25140, Washington, DC 20007.

UNICEF, *The State of the World's Children*, New York, Oxford University Press.

World Bank, *World Development Report*, New York, Oxford University Press.

World Resource Institutes, *World Resources*, New York, Oxford University Press.

Index

About the Author

Donella Meadows is a systems analyst, journalist, college professor, international coordinator of resource management institutions, and farmer. She was originally trained as a scientist, earning a bachelor's degree in chemistry from Carleton College in 1963 and a Ph.D. in biophysics from Harvard University in 1968. Since 1982 she has been teaching in the Environmental Studies Program at Dartmouth College.

From 1970 to 1972 Donella Meadows was on the team at MIT that produced the global computer model "World 3" for the Club of Rome. She is the principal author of *The Limits to Growth*, which has been translated into 28 languages and has sold millions of copies. In 1985 she began a weekly newspaper column "The Global Citizen," commenting on world events from a systems point of view. The column appears in more than twenty papers nationwide.

With Dennis Meadows she founded and coordinates the International Network of Resource Information Centers. INRIC is a coalition of systems-oriented analytical centers in twenty nations, which is working to promote sustainable, high-productivity resource management.

Donella lives on a small, communal, organic farm in New Hampshire, where she works directly at sustainable resource management.

Also Available from Island Press

Ancient Forests of the Pacific Northwest
By Elliott A. Norse

Balancing on the Brink of Extinction: The Endangered Species Act and Lessons for the Future
Edited by Kathryn A. Kohm

Better Trout Habitat: A Guide to Stream Restoration and Management
By Christopher J. Hunter

Beyond 40 Percent: Record-Setting Recycling and Composing Programs
The Institute for Local Self-Reliance

The Challenge of Global Warming
Edited by Dean Edwin Abrahamson

Coastal Alert: Ecosystems, Energy, and Offshore Oil Drilling
By Dwight Holing

The Complete Guide to Environmental Careers
The CEIP Fund

Economics of Protected Areas
By John A. Dixon and Paul B. Sherman

Environmental Agenda for the Future
Edited by Robert Cahn

Environmental Disputes: Community Involvement in Conflict Resolution
By James E. Crowfoot and Julia M. Wondolleck

Fighting Toxics: A Manual for Protecting Your Family, Community, and Workplace
Edited by Gary Cohen and John O'Connor

Forests and Forestry in China: Changing Patterns of Resource Development
By S. D. Richardson

From *The Land*
Edited and compiled by Nancy P. Pittman

Hazardous Waste from Small Quantity Generators
By Seymour I. Schwartz and Wendy B. Pratt

Holistic Resource Management Workbook
By Allan Savory

299

In Praise of Nature
Edited and with essays by Stephanie Mills

The Living Ocean: Understanding and Protecting Marine Biodiversity
By Boyce Thorne Miller and John Catena

Natural Resources for the 21st Century
Edited by R. Neil Sampson and Dwight Hair

The New York Environment Book
By Eric A. Goldstein and Mark A. Izeman

Overtapped Oasis; Reform or Revolution for Western Water
By Marc Reisner and Sarah Bates

Permaculture: A Practical Guide for a Sustainable Future
By Bill Mollison

Plastics: America's Packaging Dilemma
By Nancy Wolf and Ellen Feldman

The Poisoned Well: New Strategies for Groundwater Protection
Edited by Eric Jorgensen

Race to Save the Tropics: Ecology and Economics for a Sustainable Future
Edited by Robert Goodland

Recycling and Incineration: Evaluating the Choices
By Richard A. Denison and John Ruston

Reforming the Forest Service
By Randal O'Toole

The Rising Tide: Global Warming and World Sea Levels
By Lynne T. Edgerton

Saving the Tropical Forests
By Judith Gradwohl and Russell Greenberg

Trees, Why Do You Wait?
By Richard Critchfield

War on Waste: Can America Win Its Battle With Garbage?
By Louis Blumberg and Robert Gottlieb

Western Water Made Simple
From *High Country News*

Wetland Creation and Restoration: The Status of the Science
Edited by Mary E. Kentula and Jon A. Kusler

Wildlife and Habitats in Managed Landscapes
Edited by Jon E. Rodiek and Eric G. Bolen

For a complete catalog of Island Press publications, please write:
    Island Press
    Box 7
    Covelo, CA 95428

or call: 1-800-828-1302

Island Press Board of Directors